Context-Aware, Real-Time Fleet Management

In an emergency, every second counts, and even a small mistake can cost lives. Yet emergency fleets – ambulances, fire trucks, and police vehicles – still rely on platforms that are unable to integrate the diverse data needed to support accurate, reliable, and timely decisions. Selecting which vehicle to dispatch to an emergency is one such decision – a task that is complex, stressful, and prone to error with conventional dispatch systems.

This book examines how connected vehicle platforms – the technological backbone of modern fleet management – can be designed to meet the operational requirements of emergency services. By integrating vehicle and context data in real time – covering everything from vehicle status to staff availability, crew skills, shift schedules, traffic, and weather – these platforms enable faster and more consistent decision-making. Using the UK National Health Service ambulance fleet as a case study, the book presents a prototype system that automates ambulance dispatch, demonstrating how a thoughtfully designed platform can deliver performance and capabilities far beyond conventional fleet management systems.

Features:

- Presents a comprehensive overview of fleet management with core principles and provides essential knowledge for understanding and improving fleet management
- Highlights the value of custom-built data platforms, which enable more efficient and effective fleet management than conventional, off-the-shelf systems
- Explains how to design fleet management systems from system requirements, offering a unique engineering approach not found in other books
- Includes an end-to-end development process and covers the full process from design through testing, providing a practical, hands-on guide for creating effective fleet management solutions
- Demonstrates how real-time integration of vehicle data and context data transforms the management of the UK NHS ambulance fleet

For fleet managers, emergency service professionals, researchers, and policymakers, this book provides a roadmap to improve fleet performance, reduce operational risk, and ultimately save lives – showing why every fleet needs a data platform tailored to its unique requirements.

Context-Aware, Real-Time Fleet Management

Next-Generation Platforms for Emergency Services

Salvatore Gariuolo
Stefano Longo

CRC Press
Taylor & Francis Group
Boca Raton London New York

CRC Press is an imprint of the
Taylor & Francis Group, an **informa** business

Designed cover image: Sergio Limongelli

First edition published 2027
by CRC Press
2385 NW Executive Center Drive, Suite 320, Boca Raton FL 33431

and by CRC Press
4 Park Square, Milton Park, Abingdon, Oxon, OX14 4RN

CRC Press is an imprint of Taylor & Francis Group, LLC

ISBN: 978-1-041-17213-0 (hbk)
ISBN: 978-1-041-17214-7 (pbk)
ISBN: 978-1-003-68854-9 (ebk)

DOI: 10.1201/9781003688549

Typeset in Times
by SPi Technologies India Pvt Ltd (Straive)

Dedication

To Carla and Santiago,
To Luciano, Gaetana, and Angela.

Contents

Listings

Abbreviations

A&E	Accident and Emergency
ACID	Atomicity, Consistency, Isolation, and Durability
ADR	Australian Design Rule
AJAX	Asynchronous JavaScript and XML
AMPDS	Advanced Medical Priority Dispatch System
AMQP	Advanced Message Queuing Protocol
API	Application Programming Interface
AT	Attention
AUX	Auxiliary Input Jack
AVL	Automatic Vehicle Locator
CAD	Computer-Aided Dispatch
CARB	California Air Resources Board
CAV	Connected and Autonomous Vehicle
CI	Complexity Index
CO	Carbon monoxide
CoAP	Constrained Application Protocol
COREPER	Comité des représentants permanents
COTS	Commercial Off-the-Shelf
COVID	Coronavirus
CPU	Central Processing Unit
CSS	Cascading Style Sheets
CV	Connected Vehicle
CVP	Connected Vehicle Platform
DASH	Data Awareness for Sending Help
DB	Database
DBMS	Database Management System
DCA	Double Crewed Ambulance
DOM	Document Object Model
DTC	Diagnostic Trouble Codes
ECU	Electronic Control Unit
EMS	Emergency Medical Systems
EMT	Emergency Medical Technician
EOBD	European On-Board Diagnostics
ETI	Equipment and Tool Institute
EU	European Union
EV	Electric Vehicle
FMS	Fleet Management Systems
GIS	Geographic Information System
GNSS	Global Navigation Satellite System
GPIO	General-Purpose Input/Output
GPS	Global Positioning System
GUI	Graphical User Interface

HC	Hydrocarbons
HDMI	High-Definition Multimedia Interface
HDV	Heavy Duty Vehicle
HEV	Hybrid Electric Vehicle
HQL	Hibernate Query Language
HTTP	Hypertext Transfer Protocol
I/O	Input/Output
IAM	Institute of Asset Management
ICT	Information and Communication Technology
ID	Identifier
IEEE	Institute of Electrical and Electronics Engineers
IOS	Internet Operating System
IoT	Internet of Things
IP	Internet Protocol
ISO	International Standards Organization
IT	Information Technology
ITS	Intelligent Transportation System
JCT	Job Cycle Time
JOBD	Japanese On-Board Diagnostics
JSON	JavaScript Object Notation
LGV	Light Goods Vehicle
MAC	Media Access Control
MIL	Malfunction Indicator Light/Lamp
MQ	Message Queueing
MQTT	MQ Telemetry Transport
MVC	Model-View-Controller
MVP	Model-View-Presenter
MVVM	Model-View-ViewModel
NAEI	National Atmospheric Emission Inventory
NAT	Network Address Translation
NFR	Non-Functional Requirement
NHS	National Health Service
NoSQL	Not-only SQL
NOx	Nitrogen Oxides
OBD	On-Board Diagnostics
OEM	Original Equipment Manufacturer
OLPT	OnLine Transaction Processing
OS	Operating System
PHEM	Passenger car and Heavy-duty Emission Model
PID	Parameter ID
PM	Particulate Matter
PTS	Patient Transport Service
QoS	Quality of Service
QR	Quick Response
RAM	Random Access Memory
REST	REpresentational State Transfer

RTOS	Real-Time Operating System
RVV	Rapid Response Vehicle
SAE	Society of Automotive Engineers
SFPD	San Francisco Police Department
SMPS	Switched-Mode Power Supply
SMS	Short Message/Messaging Service
SQL	Structured Query Language
SRID	Spatial Reference Identifier
SRS	Software Requirements Specification
SSL	Secure Sockets Layer
TCP	Transmission Control Protocol
TLS	Transport Layer Security
UCA	Urgent Care Assistant
UI	User Interface
UK	United Kingdom
UML	Unified Modelling Language
URI	Uniform Resource Identifier
URL	Uniform Resource Locator
USB	Universal Serial Bus
USP	Unique Selling Point
V2X	Vehicle-to-Everything
VIN	Vehicle Identification Number
XML	eXtensible Markup Language

Preface

When I first started working with connected vehicles, I was fascinated by the sheer amount of data they generate every second. Cars, buses, and ambulances are no longer just vehicles – they are moving computers, continuously producing information about their status and performance. But I soon realised that the real opportunity goes beyond the vehicles themselves. Together, as fleets, these vehicles operate in a rich and constantly changing environment – traffic, weather, staff, and operational priorities – all shaping the decisions that must be made about the fleet. By combining vehicle and context data, fleet management can be transformed: faster, better-informed, and more reliable decisions become possible.

Imagine a life-or-death emergency. A 999 call comes in, reporting a patient in cardiac arrest. In the control room, a dispatcher must decide – within seconds – which ambulance to send. There are dozens of vehicles in the area, each with different crews, skills, and distances from the patient. Traffic is unpredictable. Roads may be closed. Every second counts. Even the most experienced dispatcher can process only so much information at once. Mistakes happen, and delays can cost lives. Yet all the data needed to make the right decision – instantly, without any delay – already exists, if only it could be gathered, organised, and used effectively.

As I looked more closely at the systems used to manage emergency fleets, I noticed an alarming gap. Despite the enormous amount of data available, many emergency services – ambulances, fire trucks, and police vehicles – still rely on conventional fleet management platforms. These platforms focus only on vehicle data, leaving out the rich context surrounding emergency operations. Without this context, they cannot make rapid, automated, or reliable decisions. Critical decisions – like choosing the best ambulance to dispatch – fall entirely on human operators, who must quickly piece together fragmented information under intense pressure, knowing that even a few seconds of delay can have serious consequences.

It was this challenge that inspired me to write this book. I wanted to understand why, despite all our technological progress, emergency fleets still depend on unsuitable systems – and how we could do better. The answer lies in context-aware, real-time fleet management: bringing together every piece of information to make instant, reliable decisions. Using the NHS ambulance service as a case study, this book shows how a well-designed data platform can transform the way fleets operate. Unlike conventional systems, such a platform integrates vehicle data and context data in real time, allowing many tasks that still depend on human judgement – like ambulance dispatch – to be automated, reducing delays and minimising errors.

Although grounded in research, this book is written for a wide audience. Whether you are a fleet manager, an emergency service professional, a researcher, or simply curious about how data and technology can improve emergency services, I hope this book provides you with insight into the challenges, opportunities, and impact of modern fleet management. More than anything, this is a story of how thoughtful technology can make fleet operations faster, smarter, and more reliable – and, in doing so, save lives.

Salvatore Gariuolo

This book represents the culmination of a journey that began several years ago, when Salvatore first joined my research group as a doctoral student and later as a Research Fellow. What started as a technical challenge in connected vehicle systems quickly became something much larger: an opportunity to rethink how data, software engineering, and transport systems can come together to support decisions that matter deeply, sometimes critically, in real-world operations.

Cranfield University has a long and distinguished history in automotive engineering, yet at the time, we were still at the early stages of engaging with the emerging challenges of vehicle connectivity, autonomy, and large-scale data integration. Salvatore's deep expertise in software engineering, paired with the department's breadth of knowledge in vehicles and systems, created an ideal environment in which new ideas could take shape. Additional support from the UK Government-funded research provided not only resources but also the freedom to explore a problem space that was both technically rich and societally important. Together, these conditions seeded the concepts that eventually flourished into the work presented in this book.

Although we have grounded our case study in the National Health Service ambulance fleet, the implications extend far beyond one organisation or sector. The question at the heart of this book, how to design decision-support platforms that integrate complex, heterogeneous, real-time data, resonates across all modern fleets, public and private. As vehicles become increasingly connected and operational environments more dynamic, the ability to combine vehicle behaviour with contextual awareness is no longer a technical curiosity; it is a prerequisite for safe, efficient, and intelligent mobility systems.

Salvatore has carried the intellectual weight of this work with dedication, curiosity, and rigour. It has been a privilege to see his ideas evolve from initial concepts, to research prototypes, to a coherent framework that can support practitioners, researchers, and policymakers alike. My contribution has been to support, challenge, and guide where useful; the achievements here are very much his own.

I hope that the insights presented in these pages will encourage others to consider what becomes possible when engineering, data, and operational realities are brought together with clarity of purpose. We are entering a period in which mobility systems will generate unprecedented amounts of data. Transforming that data into meaningful, reliable decisions will be one of the defining engineering challenges of the coming decade. This book is a step towards meeting that challenge.

Stefano Longo

Authors

Salvatore Gariuolo is a leading expert in the future of mobility and the broader impact of emerging technologies. His work bridges academic insight with practical application, helping both industry and the public navigate a rapidly evolving technology landscape. He regularly presents his work at international conferences and shares insights through podcasts and interviews.

He currently serves as senior threat researcher at Trend Micro, where he investigates risks across a wide range of technologies, from connected vehicles to artificial intelligence (AI). His research examines how these complex technologies may evolve and create new threats, informing strategies to mitigate risks before they materialise.

Gariuolo holds a PhD from Cranfield University and a master's degree in computer science engineering from the University of Naples Federico II. With nearly a decade of research experience, he previously served as research fellow at Cranfield University, leading projects on connected and autonomous vehicle systems. He is also the founder of VEICOLNET, a company advising the automotive sector on strategy, innovation, and security, helping organisations navigate emerging technologies, regulatory compliance, and risk.

Stefano Longo is an internationally recognised leader in AI-driven mobility, specialising in intelligent decision-making algorithms that power everything from advanced battery systems to autonomous driving. With a career spanning academia, industry, and deep-tech innovation, he has delivered more than ten major R&D programmes and contributed to shaping the next generation of automated and electrified transport.

He currently serves as head of Intellectual Property at Embotech AG in Switzerland, associate professor in Automated Vehicles at Cranfield University in the UK, and a strategic advisor for startups and incubation centres. His work blends cutting-edge research with real-world deployment, supported by a strong record of collaboration with OEMs, Tier-1s, and technology start-ups. Stefano is an inventor on a growing portfolio of patents and author of multiple books and over 90 publications that have influenced industrial practice and academic research in control, optimisation, and autonomous systems.

He has held senior academic and industrial roles, directed specialist postgraduate programmes, supervised extensive doctoral research, and contributed to national and international research councils. He is a senior member of IEEE, a chartered engineer, and a fellow of the Higher Education Academy and is regularly invited to speak at global conferences on autonomy, electrification, and intelligent mobility.

1 Introduction

The remarkable progress in Information and Communication Technologies (ICT) over the last few decades has paved the way for the introduction of revolutionary services that would have been inconceivable before. We are now entering an era in which the ever-expanding digital world is seamlessly integrated into our reality. We are turning into conscious yet unaware spectators of widespread, invisible interactions between our physical surroundings and the Internet: everyday objects, from cameras and traffic sensors to wearable devices, as well as home appliances, communicate their data over the network, where data can be processed to provide richer insights. McKinsey estimates that 127 new devices are being connected to the Internet every second [1], with a projection of 30 billion devices connected by 2030 [2]. We are, therefore, witnessing an explosion in the volume of data, together with a tremendous increase in its richness and diversity. This phenomenon is well captured by the increasingly popular notion of the Internet of Things (IoT), which refers to these objects as interconnected sensors and processing devices with the ability to accumulate, compute, and transfer data to a designated end system with the assistance of web technologies [3].

Virtually any everyday object can be brought under the umbrella of the IoT paradigm, and vehicles are no exception to this. Today's cars contain about 40 microprocessors, dozens of sensors, and thousands of components [4], and their complexity is only going to increase: in the future, cars are expected to record and produce an enormous amount of data, demanding up to 4,000 gigabytes of storage per day [5].

In the past 30 years, the automotive industry has also experienced unprecedented progress in vehicular communication, with cars now capable of interacting with other entities – a concept known as Vehicle-to-Everything (V2X) communication. The deployment of 5G communication, for instance, enables vehicles to exchange data rapidly with distant systems over the Internet. This means that data, which is the lifeblood of modern vehicles, can be shared with the outside world to enable previously unimaginable services. This evolution of cars into data-driven, highly connected systems is what makes them fit into the IoT paradigm; however, rather than referring to cars as 'things', we use the more familiar term Connected Vehicles (CVs). Needless to say, this paradigm represents a tremendous opportunity: with data, the automotive revenue pool could grow and diversify through new services, potentially creating a \$1.5 trillion market by 2030 [6].

However, data – in its raw and most basic form – are virtually worthless, as it consists of unorganised facts that need to be processed [7]. Only when elaborated, organised, and presented in a certain context can we draw valuable insights from data, which then becomes information. This means that the platforms that collect and process vehicle data are key to the success of a connected vehicle application. These platforms, which are commonly referred to as Connected Vehicle Platform (CVPs), constitute the fundamental infrastructure to support intelligent computing and

DOI: 10.1201/9781003688549-1

decision-making within the novel automotive IoT ecosystem. They can target different types of vehicles, such as private cars and commercial fleets, and serve a wide variety of services, such as tracking and monitoring. Although the fundamental infrastructure of a CVP always incorporates certain core components, we should not fall into the trap of assuming that these platforms are generic and can support any fleet management application: an IoT system is a very complex apparatus, composed of different elements that work harmoniously towards a common goal [8]. For this reason, the choice of the data platform proves to be one of the pillars upon which the success of connected vehicle applications should be built.

1.1 LIMITATIONS OF EXISTING DATA PLATFORMS

Context-aware, real-time fleet management applications are a particular class of connected vehicle applications that stand apart from traditional fleet management due to their peculiar traits in terms of data collection and processing. Essentially, these applications are not only based on vehicle data, but are also heavily dependent on a rich set of context data. Another requirement, lacking a trivial technological solution, is that this unusual dataset has to be processed in narrow time windows immediately after data collection. To better understand which types of fleets fall under this category of connected vehicle applications, we must explore two fundamental concepts, namely, context-awareness and real-time constraints:

- *Context-awareness* refers to the ability to use context to provide relevant information and/or services to the user. Context is practically any information that can be used to characterise the situation of an entity, whether a person, a place, or an object that is considered relevant to the interaction between a user and an application, including the user and the application itself.
- *Real-time* refers to the ability of a process to guarantee that its execution is completed within a certain time window. The correctness of a computation that is subject to real-time constraints, therefore, depends not only on its logical correctness, but also on the time required for the results to be produced. Reliability is also crucial for a real-time system, since a failure could cause an economic disaster or the loss of human lives.

A context-aware, real-time fleet management application thus refers to the management of a group of vehicles operating in a complex environment that includes other entities than the vehicles themselves. One of the key requirements of these applications is that decision-making, which is based on the complex relationships between the various entities operating in the fleet environment, must be performed immediately, as any delay might lead to an operational failure. This is particularly true for emergency vehicles, whose decision-making depends on the status of the vehicle and a high multiplicity of situational factors, such as the staff operating these vehicles and their skills, the emergencies these vehicles should serve, and even exceptional circumstances related to traffic conditions and meteorological factors. Operations involving emergency vehicles, such as routing and dispatch, are undoubtedly time-critical: the main objective is to reduce vehicle response times, as shorter response times are directly correlated with lower mortality rates.

As previously mentioned, a crucial ingredient for the success of a connected vehicle application is the CVP that collects and manages its data. In today's landscape, a variety of platforms are available for fleet management, ranging from systems that monitor driving behaviour to those that optimise fuel consumption. It is, therefore, reasonable to assume that these commercial platforms could also support context-aware, real-time fleet management applications. However, traditional CVPs are primarily designed for standard fleet management tasks, which rely only on vehicle status and location data – without accounting for the broader operational context. Furthermore, these platforms are not designed to process data in real-time or within narrow time windows. In fact, data processing in conventional fleet management applications is not expected to be performed immediately after data collection; instead, it can be carried out at a later stage to provide insights into the state of the fleet.

While commercial platforms currently fail to support context-aware, real-time applications, one might expect research to help bridge this gap. In fact, emergency fleet management is a widely discussed topic in academia, with much of the focus placed on developing novel algorithms for vehicle routing, dispatching, and redeployment. Yet, these studies systematically overlook the underlying framework for data collection, computing, and presentation. They often assume that such algorithms can always run on the most up-to-date data, with performance determined solely by the algorithm itself. This assumption is unrealistic, as the performance of these algorithms is significantly influenced by the underlying CVP, which dictates both the speed at which the necessary data become available and the efficiency of its processing. In other words, many academic studies often rest on a flawed premise – they implicitly assume the existence of commercial platforms capable of supporting these algorithms in real-world scenarios, when, in reality, such platforms do not exist.

As observed, the academic literature seems to overlook the technological characteristics of the CVP, which are, instead, critical for the real operating conditions of applications like context-aware, real-time fleet management. Meanwhile, no commercial platforms currently meet the unusual requirements of these applications. Arguably, this contradiction has so far limited the proliferation of appropriate solutions for emergency vehicles. However, we cannot rule out the possibility that these applications simply do not represent a significant business opportunity, or at least one that is substantial enough to justify a major departure from the current design of fleet management platforms. Irrespective of this, we find ourselves in an alarming situation: the management of certain classes of vehicles, such as emergency fleets, continues to be carried out with unsuitable data platforms, leading to inefficiencies, financial losses, and – regrettably – the loss of human lives.

1.2 RETHINKING DATA PLATFORMS FOR EMERGENCY FLEETS

The aim of our research is to design and develop a CVP capable of supporting context-aware, real-time fleet management applications, enabling performance and services that – while conceptually possible – are far from being implemented and validated in practice. Our main objective is to improve the management of critical classes of road vehicles, such as emergency fleets, by enabling real-time, automated decision-making across their operations. This is achieved through a reasoned selection of the technologies for the CVP – technologies that, together, enable an innovative approach to data management, making it possible to access and capitalise on information that would otherwise be unavailable.

This book, which summarises the findings of our research, aims to:

- **Reorganise the existing knowledge on fleet management**, which appears to be extremely fragmented and incomplete, since it primarily originates from the commercial sector.
- **Define the characteristics that a context-aware, real-time fleet management application must possess to qualify as such**, and summarise the data management requirements that arise from these characteristics.
- **Derive the fundamental design of a CVP capable of effectively supporting context-aware, real-time fleet management applications**, ensuring efficient collection, storage, and interrogation of their unique dataset.
- **Build a prototype for managing the ambulance service of the English NHS**, demonstrating the advantages of our platform over traditional Fleet Management Systems (FMS).

The novelty of our study lies in the fact that it approaches fleet management from a completely new perspective. This research is based on the central hypothesis that data management is one of the key factors determining the actual performance of a fleet management application. This means that, unless the CVP enables proper handling of the application data, the performance of the services built upon it can never be satisfactory. Yet, such concerns seem to be invariably overlooked both in academic research and by the companies developing fleet management platforms.

This oversight is particularly evident in research studies. Although the concept of fleet management originates from the commercial sector, the academic literature is teeming with studies on fleet management applications, primarily focused on developing innovative algorithms for fleet operations, such as vehicle allocation, dispatching, and route management. While this research may be innovative, it proves to be flawed, since the analysis of these solutions systematically stops at simulation. This implies that all practical factors that could influence algorithm performance are inevitably overlooked.

In most cases, this is not a concern, as commercial data platforms can adequately handle the necessary application data, ensuring no significant degradation in performance when transitioning from theory to practice. However, certain fleets require specialised data management, imposing additional considerations on the system handling their data, which are essential to ensure that algorithm performance remains satisfactory under real-world conditions. Emergency fleets, which rely on context-aware, real-time data management, provide a clear example. Using the ambulance fleet as a case study, this work proves that rethinking fleet management applications – particularly their data management requirements – is often necessary to turn the pioneering ideas of the academic world into reality.

The significance of this research is further evidenced by its application in practical projects. For example, the OBD-II dongle developed as part of this research was used in Project Alloyed, a collaborative initiative funded by Innovate UK to advance innovations in vehicle connectivity. This work also led to participation in two academic projects with a strong commercial focus. The first, under the ICURe pre-acceleration programme, explored market opportunities in the automotive sector arising from the use of context data. The second, within the MICRA programme, sought to identify fleet applications requiring bespoke data management. These outcomes highlight the relevance of our research, not only in advancing scientific knowledge but also in addressing real-world challenges.

It is worth noting that, with recent advances in artificial intelligence (AI), the concepts presented in this book are more relevant than ever. While these technologies promise powerful decision-making capabilities, they cannot function without a solid foundation of high-quality, timely, and context-rich data. The CVP described in this book provides exactly that: a reliable architecture for collecting, storing, and processing diverse datasets in real time. Whether fleet management decisions are made using traditional algorithms or AI-driven approaches, the ability to access the right information at the right moment remains indispensable. In this sense, the principles and methods presented here form a necessary precondition for any fleet management solution.

1.3 STRUCTURE OF THE BOOK

The remainder of this book is organised as follows:

- **Chapter 2** reorganises the existing knowledge on fleet management by reviewing both academic literature and the available commercial solutions. This chapter introduces the definition of fleet management and outlines its key characteristics. It then focuses on context-aware, real-time applications, their requirements, and the challenges that must be overcome to ensure effective fleet management.
- **Chapter 3** outlines the requirements for the efficient management of the English NHS ambulance fleet, which serves as a case study for implementing the CVP. Based on these requirements, and using modern software engineering methods, the fundamental design of the platform is derived. The individual components of the CVP are presented in the subsequent chapters.
- **Chapter 4** focuses on the design and development of the data storage infrastructure, which proves to be the cornerstone of the CVP. This chapter demonstrates how this initial design choice is the factor behind the success or failure of the platform, and discusses the need to break with the database technologies used in conventional fleet management systems.
- **Chapter 5** describes the steps necessary to implement a communication interface that guarantees responsive database ingestion of both vehicle data and context data. This process also considers additional requirements, such as security and reliability, which are crucial for these novel IoT applications.

- **Chapter 6** details the process of building a data-gathering component that turns any vehicle into an IoT device. Specifically, it explains the rationale behind the development of a next-generation On-Board Diagnostic (OBD) dongle, which significantly deviates – in terms of both structure and performance – from the traditional telematics devices used for fleet management.
- **Chapter 7** focuses on the design and development of the user interfaces for the CVP. The first section covers the Android application, which allows ambulance operators to interact with the rest of the platform via their mobile phones; this application serves both as a primary source of context data and as a distributed interface to consume the data processing output. The second section is dedicated to the web app, which allows end users to monitor fleet data in real-time and initiate the automated ambulance dispatching service.
- **Chapter 8** analyses the functioning of the CVP by testing the behaviour of the ambulance dispatching procedure. The chapter demonstrates how our data platform meets the unique requirements of a context-aware, real-time fleet management application, and highlights the performance advantages that the CVP offers over traditional dispatching systems.
- **Chapter 9** explores the significance of our findings and discusses the challenges that remain to be addressed.

2 Overview of the State-of-the-Art

Although the definition of context-aware, real-time fleet management is introduced for the first time in this book, such applications are far from novel in academic research. Proof of this is the extensive literature on emergency fleet management, which often focuses on critical operations such as routing and dispatch – tasks that must be finely tuned to achieve optimal results, as any delay or error could lead to disastrous consequences. This usually translates into the search for efficient algorithms to solve complex optimisation problems, which require sophisticated and performance-intensive mathematical computations. However, even when research provides answers to these complex problems, not all solutions can be implemented in practice. In today's commercial landscape, no CVP is capable of efficiently running these algorithms, as their data processing requirements differ significantly from typical FMSs.

The main objective of this chapter is to demonstrate, through an extensive review of the academic literature, that virtually any research on context-aware, real-time fleet applications overlooks the technological architecture for data management, posing a significant barrier to the rapid and effective adoption of such solutions. The section on fleet management introduces its definition, its typical applications, and the main characteristics of the underlying data management platform. The focus then shifts to emergency fleet applications, which are the most representative example of context-aware, real-time fleet management. The next section examines how the academic literature approaches the research questions related to the management of emergency vehicles; it emphasises the need to adopt new data management platforms, distinct from those currently on the market, to support the unique requirements of these applications.

2.1 FLEET MANAGEMENT

Nowadays, many companies rely on transport for their business: the performance of these companies, and in turn the satisfaction of their customers, goes hand in hand with the efficiency of their vehicle operations. Keeping fleet expenses under control is equally important, as fuel, maintenance, and operational costs can easily spiral out of control. Avoidable expenses should be identified and eliminated to ensure cost-effectiveness. This has led to the emergence of FMS, designed to reduce fleet costs, improve operational efficiency, and increase the visibility of vehicle activities.

This section outlines the main characteristics of traditional fleet management, providing a foundation for defining context-aware, real-time fleet management

DOI: 10.1201/9781003688549-2

applications. It also identifies the technological architecture that underpins existing commercial FMSs, which is shaped by the applications these platforms are designed to support.

2.1.1 Definition and Characteristics

The notion of fleet management has slipped unobtrusively into common usage without much definition: it is frequently used to describe a variety of commercial products and services that companies can adopt to improve the efficiency of their vehicles. Each service provider may therefore adopt a different definition, depending on its proposition, business model, or even advertising campaign. This, however, creates considerable confusion over the true meaning of the term. To derive an unambiguous definition of fleet management and the fundamental characteristics of its applications, it is therefore necessary to conduct an analysis of the term's meaning, which begins with understanding the role of vehicle fleets in the modern corporate world.

Vehicle fleets are groups of motor vehicles owned or leased by companies and organisations, rather than individuals and families, to support their operations. Typical examples include fleets operated by public bodies, transport operators, police forces, ambulance trusts, taxi companies, and car rental firms. Trucks and lorries used for delivering goods also fall under this category. In some countries, even vehicles privately owned by employees but used for work purposes – known as grey fleets – are legally recognised as such. Fleet sizes vary widely, from small-business fleets with just a few cars to gargantuan groups, such as the UK vehicle leasing company Lex Autolease, which operates 283,000 vehicles and is regarded as the largest fleet in the world. For many companies, regardless of fleet size, vehicles are a key resource that requires appropriate management like any other company asset. This means that the definition of fleet management can be derived from that of asset management.

According to ISO 55000, asset management is a coordinated activity of an organisation to realise value from its assets. A more intuitive way to understand this concept is through conceptual models, which graphically describe the content of the standard. Figure 2.1, for example, shows the conceptual model developed by the Institute of Asset Management (IAM). As illustrated in the figure, asset management can be defined as a systematic process of coordinating finances, operations, risks, and all other asset-related entities to increase the value of a company's assets. This is achieved by acting, in the most cost-effective way possible, on the entire life cycle of the assets – from planning and development, through maintenance and renewal, up to their disposal.

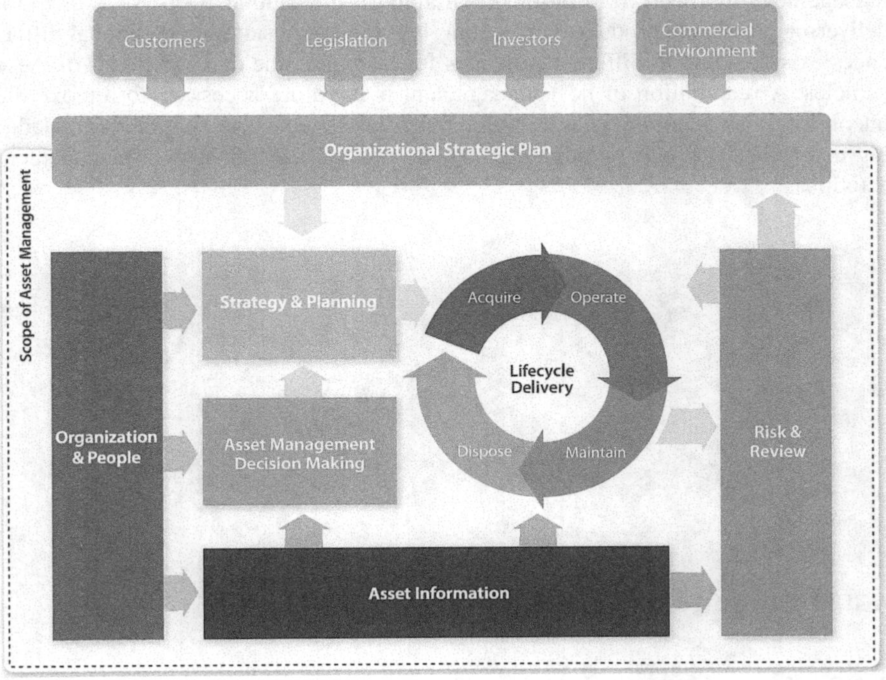

FIGURE 2.1 IAM conceptual model for asset management.

Considering the vehicle fleet as a corporate asset and drawing on the definition of asset management provided by ISO 55000, fleet management can then be defined as a group of activities that enable companies whose operations depend on transport to minimise the risks associated with their vehicle strategy. The primary objective of these activities is to reduce overall transport costs – including staff-related expenses – while ensuring adequate performance and full compliance with governmental legislation and any other constraints that might be in place.

From the definition we have derived, fleet management should therefore refer to the entire decision-making process that transforms any available information into informed decisions concerning the life cycle of fleet vehicles. Influenced by the many commercial value propositions around us, we often misconstrue this definition, restricting fleet management to a small set of activities that use vehicle data to monitor and reduce fleet-related costs in the short term. SIMFORM's value proposition, shown in Figure 2.2, provides an example of this, as it restricts fleet management to a few activities such as fuel management and driving behaviour monitoring. In reality, fleet management activities are a part of a complex strategic process that depends on both the business and operational contexts within which the vehicles operate, as well as on their functional objectives, expected performance, and the stakeholders involved. Consequently, it is not possible to reduce all fleet management activities to a small set that works for every fleet. Ambulances, for example, require a totally different management approach from buses. Even among very similar fleets, such as those delivering perishable products versus long-life goods, substantial operational differences exist, leading to different strategies for managing the daily activities of these vehicles. A redefinition of fleet management is therefore necessary to remove the unconscious conditioning caused by the marketplace; this also entails acknowledging that each fleet must be supported by its own tailored set of fleet management activities.

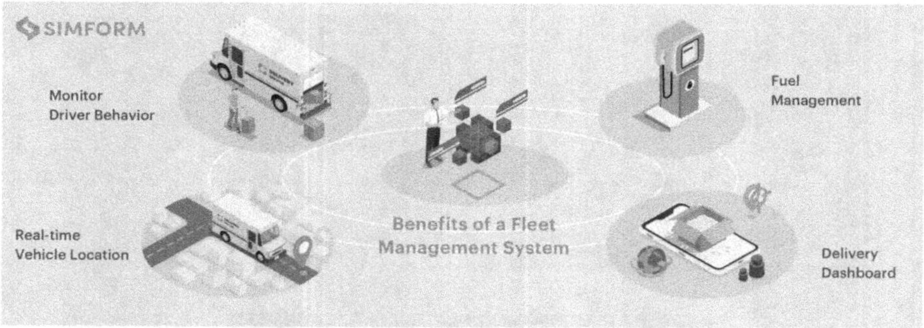

FIGURE 2.2 SIMFORM fleet management proposition.

2.1.2 Applications and Impact on Services

The definition of fleet management is broad enough to encompass a multitude of activities with different purposes and impacts on services. However, from a purely commercial perspective, fleet management activities are often limited to a small, well-defined set. This is because most of these activities are designed for a predominant market segment: commercial fleets, such as delivery vehicles. These fleets essentially require (i) assessment and optimisation of vehicle performance, (ii) live vehicle monitoring, and (iii) planning of delivery routes. This minimal set of activities, together with the smaller market segment of passenger vehicle management, represents the bulk of a market projected to reach $42.75 billion globally by 2028. A summary of the elements that conventional fleet management activities allow for control, along with the information they produce, is provided in Table 2.1.

TABLE 2.1
Fleet Management Controls and Reports

Control	Reports
Assets (vehicles, equipment, tyres, etc.)	Individual assessment of equipment
Legal obligations	Assessment of equipment groups
Staff performance assessment (drivers)	Equipment performance
Maintenance (planned maintenance)	Min and max fuel consumption averages
Tires	Exceptions vision (out of the average)
Fuel	Cost and performance parameters
Warranties	Preventative maintenance to be carried out
Services performance	Corrective maintenance
Requisition of parts	Actual cost assessment per piece of equipment
Purchasing and application of parts	Km/tire and per piece of equipment, etc.
Costs per branch, per vehicle, etc.	Records of tires and equipment
Parts and inventory	General records

By contrast, solutions capable of supporting fleets with other functional and operational requirements remain largely absent from the market. This is despite the fact that each fleet should ideally be served with a dedicated fleet management solution. For example, strategies used to assess the performance of delivery vehicles are generally unsuitable for an emergency fleet: while the former rely primarily on vehicle data, the latter should combine this data with staff and emergency information to be effective. These differences become even more pronounced for operations such as routing and dispatching. Undoubtedly, all vehicles still benefit from traditional fleet management to some extent; however, the results achieved with conventional management are minimal compared to those that a tailored fleet management solution could provide.

Stepping back from a purely commercial standpoint, a more nuanced scenario of fleet management applications emerges. Here, however, another problem arises: since the definition of fleet management is relatively broad, even the academic literature is often uncertain about how to classify its applications. A representative example is the extensive analysis by Vivaldini, Pires, and de Souza on logistics management, which considers fleet monitoring as a parallel activity to fleet management. As explained in Section 2.1, this approach is fundamentally incorrect, since all activities that use information to derive informed decisions about vehicles – such as monitoring – should be considered part of fleet management. Moreover, most research focuses on a single type of fleet, and studies analysing different aspects of the same fleet are also common. This leads to a fragmented – and sometimes confusing – view of fleet management applications.

To bring some order to this landscape, we propose grouping fleet management applications into three fundamental categories: (i) performance evaluation, (ii) monitoring, and (iii) vehicle operations management – similar to what we have done for delivery vehicles. However, this categorisation should be regarded as a reference framework rather than a strict classification, since these applications can assume different nuances depending on the vehicle fleet. The objectives, structure, and desired impact of these activities depend, in fact, on two interdependent factors: (a) the characteristics of the fleet, such as its composition, and (b) the context in which the application is deployed, which defines the target performance of the vehicles. This means that different fleets – depending on their operational objectives – require tailored approaches to managing vehicle operations.

Academic literature reflects this diversity, with studies examining a wide range of applications across different fleet types. Hoff, for example, focuses on road freight transport, using factors such as fleet composition and routing criteria to derive optimal decisions on non-predetermined trips. Cheung and Hang analyse the problem of driver-task allocation in the context of container movement within time windows and uncertainty in task duration. Coslovich, Pesenti, and Ukovich examine the same problem for container transport, where the objective is to minimise the total cost while satisfying time-window constraints. Peters, Matta, and Boe study how to manage the assignment of staff with different skills on a heterogeneous fleet of buses. The study by Li and Fu aims to minimise the costs of a bus fleet while reducing transport time for students. By modifying single elements of the operating environment of the same fleet, completely different problems can arise: Park, Tae, and Lim analyse bus fleet management when several schools share the same bus, while Lima et al. assume, instead, a common departure time for all schools. A decision support system for improving the responsiveness of police vehicles to incidents – taking into account information such as traffic conditions and demand coverage – is developed by S. Dunnet, while an event-based system for the dynamic assignment and relocation of ambulances is designed by H. Billhardt et al. Other notable examples include car rental logistics management, analysed by Fink and Reiners and by Oliveira and Toledo, as well as waste vehicles management, investigated by Varone and Bierlaire.

Taken together, these studies show that, for any fleet type, each vehicle operations management application can be formulated as an optimisation problem. Some of these problems, defined in simpler operational contexts, rely solely on vehicle data; others, functionally more complex, also require context data. Some problems are defined in a static environment, others in a dynamic environment; some use deterministic models, others stochastic ones; and each of these optimisation problems, subject to specific constraints, uses a different cost function. However, there is one point that ties these studies together: none takes into account the characteristics of the underlying platform for data collection, processing, and visualisation. As a result, it is often impossible to effectively manage such applications in real-life settings, as will be discussed in Section 2.2. This issue has been addressed in several studies, such as that carried out by R. Salazar-Cabrera et al., which highlights the limited attention given by academic research to ITS frameworks for developing new fleet management solutions. Consequently, many fleet management applications – beyond those already on the market – remain, for the moment, confined to the academic literature, which mainly focuses on the search for effective algorithms to solve the aforementioned optimisation problems. These algorithms, however, would require entirely new CVPs.

Today, the common practice is to use traditional fleet management applications across all fleets. While these applications certainly have an impact – improving fleet performance and reducing costs – they are limited in scope. Developing new CVPs can bridge the gap between academic research and commercial practice, facilitating the adoption of more advanced fleet management solutions and enabling these applications to reach their full potential.

2.1.3 Data Platforms for Fleet Management

The CVP is the framework responsible for gathering, storing, and supplying fleet-related data. This technological infrastructure is therefore essential for any fleet management application, as it ensures that data – once cleansed, enriched, and processed – can support companies and organisations in optimising their vehicle strategies.

As discussed in the previous section, each fleet has unique characteristics and operates in a distinct environment. These differences inevitably determine the type of fleet management applications required, as each fleet faces its own operational challenges and data needs. In turn, these applications shape the structure of the underlying CVP, since the way data must be collected, processed, and used varies from one application to another. To meet these requirements effectively, the design and capabilities of the CVP must therefore be aligned with the specific needs of both the fleet and the applications it supports. When this alignment is missing – that is, when an FMS is not suited to the fleet it serves – data may not be collected accurately, processed correctly, or delivered on time, limiting the effectiveness of fleet management decisions. As such, the CVP influences every aspect of fleet management far more than the academic literature suggests; yet it remains consistently overlooked.

Because CVPs play a central role in fleet management, understanding their fundamental structure is crucial. In this section, we outline the common framework of the CVP, as well as the design criteria that enable the data platform to be adapted to the specific needs of different applications.

Even though CVPs must differ in their structure to properly support various fleet management applications, a common denominator can always be identified. The rationale for a shared layout underlying every FMS is that any fleet management application – regardless of fleet type or operational context – requires three fundamental functionalities: (a) the gathering of fleet-related data, (b) the storage of this data in a repository, and (c) the visualisation of either live data or the results of subsequent processing. The architecture common to all FMSs is illustrated in Figure 2.3.

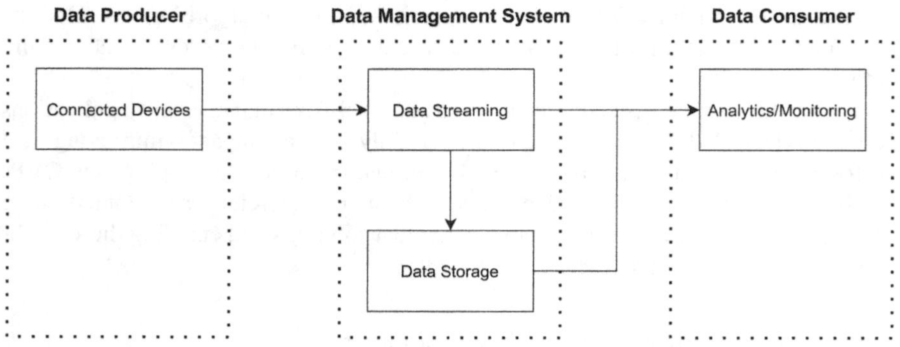

FIGURE 2.3 FMS fundamental architecture.

The result is a data platform consisting of three fundamental components:

- **Data producer**, responsible for collecting the data required by fleet management applications. For most applications, this component comes down to a set of dongles equipped with GPS and an OBD-II interface, which transmit vehicle diagnostic data over the cellular network.
- **Data management system**, responsible for handling the data collected by the data producer. Once received, this data is either transmitted to the data consumer using streaming technologies such as Apache Kafka or stored in an electronic Database (DB), which is typically relational or NoSQL.
- **Data consumer**, responsible for processing the data collected and presenting it to the end-users. The algorithms for data processing, as well as the layout and the functionalities of the user interface – usually implemented as a desktop, mobile, or web application – are a function of the chosen application.

In light of the above, we can see that the structural layout of the CVP is common to all fleet management applications. At the same time, certain characteristics of the data platform depend on the type of fleet and the operational setting in which it operates; other characteristics, in contrast, are determined by the specific application the CVP is designed to support. To fully understand the logic behind the relationship between the structure of the CVP and the application it serves, it is useful to refer to the definition of *software requirement*. According to the IEEE Standard Glossary of Software Engineering Terminology, a software requirement is defined as: (i) a condition or capability needed by a user to solve a problem or achieve an objective, (ii) a condition or capability that must be met or possessed by a system or system component to satisfy a contract, standard, specification, or other formally imposed document, and (iii) a documented representation of a condition or capability as in (i) or (ii). Software requirements, therefore, describe how a system should behave, appear, or perform. They can be used by different stakeholders to describe a system at various levels of detail. A primary classification of these requirements is shown in Figure 2.4.

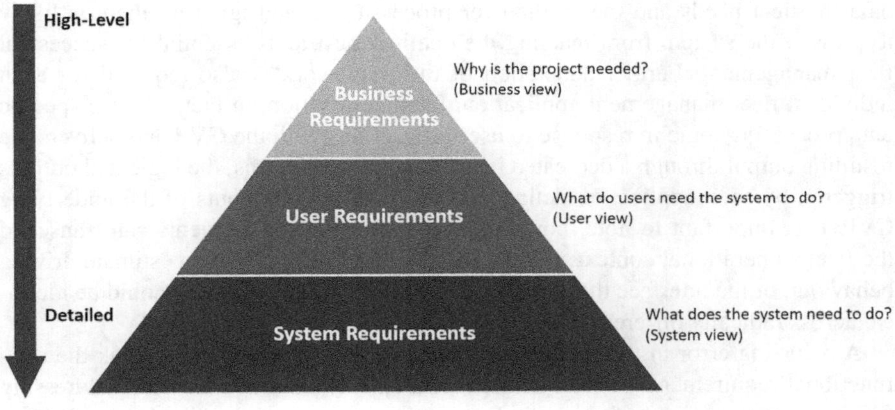

FIGURE 2.4 Product requirements hierarchy.

What links the characteristics of a fleet management application to the structure of the CVP designed to support it are the *system requirements*. These requirements are used by all stakeholders involved in the design and development process, and can be grouped into:

- **Domain requirements**, which describe the core features that any system within a given domain must necessarily exhibit. These features are characteristics of the class or segment to which the system belongs – rather than user- or application-specific – and influence the fundamental layout and main components of the system.
- **Non-functional requirements**, which define the quality constraints that a system must satisfy in order to operate effectively across a certain group of applications. They are typically classified into performance, operational, life cycle, and economic requirements, and address aspects such as performance, security, scalability, reusability, portability, and maintainability. These constraints are therefore subordinate to the context in which the system operates.
- **Functional requirements**, which specify the functions that a system must provide to meet the explicit needs of its users. These functions – usually expressed as input, operations, and expected outputs – are user-specific and primarily influence the structure of the dataset, the processing logic, and the design of the user interfaces.

All CVPs used for fleet management feature the same fundamental structure because they share a common set of domain requirements. In fact, all applications require standard functionalities, as they belong to the same industry vertical, which results in a single framework for their management. This framework, however, must be adapted to the needs of specific fleets, since each CVP – considering both the fleet to be managed and the operational environment in which its vehicles operate – has its own set of non-functional requirements. These requirements define the type of data the fleet needs and the method for processing it, and ignoring them is likely to prevent the system from meeting the quality constraints essential for successful fleet management. Further adaptation of the framework is also required for each individual fleet management application. Each application, in fact, triggers specific data processing logic in response to user interactions with the CVP and delivers the resulting output through a dedicated interface. In formal terms, the logic and outputs triggered by each application define the functional requirements of the underlying CVP. It is important to note, however, that functional requirements can transcend the fleet's operational context; for instance, the algorithms used to estimate driving behaviour, or the interface through which their results are displayed, could be identical across radically different fleets.

A recurring error in the academic literature is the tendency to overlook the non-functional requirements of fleet management applications. Most studies addressing fleet management problems have, in fact, focused exclusively on algorithms for data processing. This means investigating the functional requirements of the application while assuming the existence of a generic framework that supports any computation. This cannot be considered a minor oversight, given the consequences this inattention would have for the practical deployment of fleet management applications, especially the safety-critical ones. Nor can it be interpreted as a deliberate choice, considering that no help is forthcoming from the commercial world in this regard. Traditional FMSs are in fact designed for delivery vehicles, and therefore target a very specific set of non-functional requirements. As a result, most of the fleet management applications investigated in academic research could not be effectively supported by existing commercial systems.

2.2 CONTEXT-AWARE, REAL-TIME APPLICATIONS

Fleets ease our day-to-day activities in countless ways, and some of them – more than others – are literally vital. Such is the case with emergency fleets, such as those operated by law enforcement agencies and fire brigades, or those involved in rescue operations and used for medical purposes, which come into play when accidents occur. Our lives, particularly in the unfortunate event of an emergency, are inextricably linked to the reliability of these services, and therefore to the responsiveness of their interventions and the performance of their vehicles. Diligent management of this certainly not marginal category of fleet is hence imperative. Unfortunately, achieving this is not so straightforward, as such fleets require management that is based on situational information and subject to tight time constraints. Emergency vehicles are, in fact, the most representative – though not the only – example of fleets that require context-aware, real-time management. The purpose of this section is to introduce the key features of this particular form of fleet management, which are then used to derive the non-functional requirements of the CVP that underpins its applications. The fundamental motivation for this work is also substantiated with evidence, by revealing how these requirements are surprisingly neglected in nearly all studies on emergency vehicle management.

2.2.1 Context-Awareness

Humans are quite successful at conveying ideas. This is due not only to the richness of our language but also to an implicit understanding of everyday situations. In other words, we are able to use implied situational information – or *context* – to expand our conversational bandwidth when communicating with one another. However, this ability largely vanishes when we interact with computers, as they are not natively capable of exploiting the context of this dialogue. Advances in artificial intelligence have begun to mitigate this limitation, yet computers still lack a genuine understanding of situational context in the human sense. By enabling computers to access context, we can enrich human-computer communications and, in turn, create more sophisticated and potentially more effective computational services.

To enable context-aware computing in a system, an alteration of its structure is required. A clear understanding of what context-awareness actually represents is a prerequisite for designers to effectively amend the system: they must be able to determine what information constitutes context for the context-aware behaviour to be implemented, as this affects both the choice of the application data sources and the way this rich dataset is stored. The first attempt to define the notion of context-awareness dates back to 1994, in the work of Schilit and Theimer, who refer to context as the position and identity of nearby people and objects, as well as their changes. This definition, however, is overly specific, which makes its application particularly difficult: it provides no explanation on how to determine whether a piece of information that is not listed in this definition can be considered context. Indeed, it is impossible to enumerate which aspects should be referred to as context, as these change from situation to situation. There have been other attempts to define context, but they fall into the diametrically opposed error of providing synonyms for it. These definitions, which refer to context as the environment or situation, are in fact so generic as to be of no practical use whatsoever – just as overly specific definitions. An operational definition of context is provided by Anind K. Dey, which was already cited in Section 1.1 and to which the reader is referred. This definition provides a rather simple yet effective mechanism to enumerate contextual information for a given application: if a piece of information characterises the situation or the participants of a certain human-computer interaction, it qualifies as context. A system capable of using context to provide relevant information and/or services to the user is then classified as context-aware.

The use of this information – often untapped in fleet management – unfolds a new world of possibilities for many applications. A context-aware system can, in fact, support new features such as: (i) tagging of context to existing information to support more sophisticated analyses, (ii) presenting additional information relevant to users, and (iii) automatically executing services. All of these features are made possible by the enriched dataset consisting of contextualised vehicle data, which can therefore be referred to as *situation data*, as it describes the states of all entities relevant to the application at a particular moment.

Taking emergency vehicle fleet management as an example, context is any information that can improve the quality of the emergency services from the end-users' perspective and increase their performance from the fleet manager's point of view. As is already the case in traditional fleet management, vehicle data can be contextualised through the use of GPS (Global Positioning System) technology, which provides information on the location of tracked assets, enabling fleet managers to monitor and control drivers' behaviour, and protect vehicles via geofencing. For emergency vehicles, it is likewise advisable to contextualise the status of the vehicles in terms of the operations they are involved in – for example, by handing information on whether they are currently available to serve a call or already engaged. Equally important is the status of the staff that operates the emergency vehicles, to determine whether they are active on the vehicles at any given time and what their responsibilities are. Other situational information that may enhance emergency vehicle operations includes traffic situations, route status, and even weather conditions. The combined use of vehicle data and context data paves the way for a new level of fleet management strategies. Interestingly, the new categories of features enabled by context, outlined above, align closely with the classification of fleet management applications suggested in Section 2.1.2.

Contextualisation of vehicle data can significantly enhance the techniques used to assess the performance of emergency fleets. Context tags (i), for example, can be used to optimise the allocation of emergency vehicles across the territory based on incoming call history. By linking each vehicle's location, status, and historical call patterns, the system can identify areas of high demand and dynamically adjust vehicle positioning, thereby improving fleet response times. Similarly, context tags can help determine whether it is opportune to increase the presence of staff with specific skill sets to better serve classes of emergency interventions that are predominant in a certain area. In both cases, the use of contextual information is essential: without it, the system could not evaluate vehicle availability, staff capabilities, or operational patterns to make informed decisions. Context tags can also be leveraged to adjust the number of spare parts stored for maintenance by analysing the correlation between the urgency of calls and the driving style adopted on the vehicles deployed to serve them, as more extreme driving generally results in a higher rate of vehicle servicing. Contextualised vehicle data can also be presented to the fleet manager (ii), providing a comprehensive understanding of fleet status through live monitoring. This allows fleet managers to make better-informed operational decisions, respond quickly to unexpected events, and gain insights that are only accessible when contextual information is integrated with vehicle data. However, the application category in which context-awareness would have the greatest impact is vehicle operations management. Context, in fact, enables full automation of the decision-making process behind vehicle operations (iii), like ambulance dispatch. By combining vehicle status, location, staff availability, and operational context, the system can automatically prioritise tasks, allocate resources, and manage operations – a level of automation that relies entirely on the availability and integration of contextual information.

Existing FMSs do not contextualise vehicle data beyond positioning. As contextual information is currently unavailable, any decisions affecting emergency vehicle operations – such as dispatching – cannot be processed by automated decision-makers, and are still left to human operators. This results in decisions that are neither systematic nor instantaneous, despite the human lives at stake. Indeed, operators inevitably introduce delays and errors due to the lack of a comprehensive view of the situation and the extreme time pressure they face. By contrast, the use of context data can enable automatic, near-instantaneous, and error-free decision-making. These examples do nothing but emphasise the importance of developing CVPs with non-functional requirements that differ from those of commercial FMSs, particularly those related to context-awareness.

2.2.2 Real-Time Constraints

One of the main criteria for evaluating the outcome of an activity is *correctness*, which is often defined as the freedom from errors. In other cases, however, correctness should rather be interpreted as the degree of conformity with expectations. The term *correctness* naturally lends itself to various shades of meaning, which are, however, usually negligible. Per contra, this subtle undertone becomes significant when correctness depends not only on the result of a task but also on the time required to obtain it. Systems that require this interpretation of correctness are referred to as real-time systems. The introduction of real-time systems primarily stems from the pressing need for effective control of safety-critical activities; they are in fact conceived to manage tasks whose result must be guaranteed within a certain time limit, as exceeding these deadlines could lead to catastrophic consequences.

The above description can be formally summarised in three main components, whose interplay characterises any real-time system, as outlined in the '94 paper by K. G. Shin and P. Ramanathan. First, the most valuable resource that real-time systems must manage is time, as each task must be completed before its deadline. It follows that the timely communication of any information relevant to task completion is also essential; the correctness of the tasks that characterise such systems thus depends on both logical correctness and the time at which the result is obtained. Second, reliability is pivotal, as failure to meet task deadlines could lead to economic disasters or loss of human lives. Third, the environment must be considered an active component of the real-time system. Although seemingly negligible, this third component is one of the cornerstones on which this research is built, as we shall see in the rest of the chapter.

The aforementioned article also explains how to characterise the tasks of a real-time application according to both their arrival time and the system's tolerance for missed deadlines. Real-time application tasks can, in fact, be classified as *periodic* if they are activated at regular intervals, or *aperiodic*, otherwise. Regarding the effects of missed deadlines on the system, a distinction can be drawn between hard, firm, and soft real-time tasks. *Hard real-time* tasks are those for which any missed deadline is considered a system failure; such tasks are typical of mission-critical systems, where failing to meet time constraints can result in the loss of life or property. *Firm real-time* tasks, by contrast, allow the system to tolerate infrequent missed deadlines; a system with such tasks can survive task failures as long as they are adequately spaced, although the result produced by those tasks ceases to be useful. *Soft real-time* tasks permit frequent missed deadlines; however, the utility of the results decreases over time after the deadline expires.

A precise characterisation of the real-time application tasks is crucial, as it directly impacts the design of the system intended to handle them. To ensure that all tasks are completed within the required time constraints, it is typically necessary to intervene on the system architecture on two different levels: the node level and the system level. At the node level, each component of the system must provide both speed and predictability in task execution and in the interaction with the external world. This is often challenging, as modern systems are built using commercial-off-the-shelf (COTS) microprocessors, which do not allow complete control over elements such as instruction execution, context switching, and memory access. The adoption of a real-time operating system (RTOS) can definitely improve system predictability and determinism in task management. At the system level, the application can be considered as a set of distributed nodes interconnected through a network that communicates with the external world. Although the individual nodes may be off-the-shelf products, the system must still be custom-designed to provide the specialised support required by its applications. In other words, data procedures must be properly configured to ensure that the system can process information quickly enough to produce the expected results. This requires an adequate data acquisition process, as a real-time system cannot process data faster than it can be acquired from its sources. A suitable communication network is also essential: regardless of its topology, the network should guarantee high speed, performance scalability, reliability, and security.

Several emergency vehicle operations can be classified as real-time applications, as they must be completed within a limited time frame to prevent potentially hazardous consequences. A prime example is ambulance dispatch, a macro-activity consisting of several interdependent tasks. It can itself be regarded as an aperiodic task, activated when the emergency call is answered by the call-taker. Although not immediately obvious – given the tremendous consequences that a delayed ambulance arrival may entail – dispatch should be modelled not as hard real-time task, but as a soft real-time one: the maximum response time, which depends on the urgency of the call, is fixed, and the utility of its result – the patient's chance of survival – does not cease after the deadline but gradually decreases. Regrettably, publicly available productivity metrics indicate the inadequate management of ambulance dispatch: in July 2018, eight out of ten NHS trusts failed to meet both the 7-minute life-threatening call response standard and the 18-minute emergency call mean response time. This gap between expected and actual performance plainly emphasises the importance of upgrading the CVP used for emergency fleet management, which is crucial to bridge this discrepancy.

2.2.3 Non-Functional Requirements and CVP Structure

The central claim that has emerged from this chapter is that the key structural features of a system are dictated by the applications it has to manage. This principle explains why platforms typically used for traditional fleet management prove unsuitable for other types of fleet applications. Some of these applications, in fact, require a bespoke CVP due to their peculiar requirements, which have a significant impact on the system's structure.

The objective of this section is to summarise the non-functional requirements of context-aware, real-time fleet management applications, and to derive from them the fundamental layout of the CVP intended to support these applications. Undoubtedly, context-aware, real-time applications are characterised by a multitude of requirements: this book does not aim to provide an exhaustive list, but rather to present a clear outline of those requirements that make their data platform unique. Finally, the structural limitations of existing FMSs in handling the unusual non-functional requirements of the context-aware, real-time fleet management applications are highlighted.

Context-aware, real-time fleet management is highly dependent on situational data: vehicle data must be contextualised because the state of the vehicles depends on a multitude of external factors, as these fleets operate in an environment that includes entities beyond the vehicles themselves. The first non-functional requirement (NFR) of the target applications can therefore be summarised as follows:

(NFR1) The CVP shall collect any information that is relevant to describing the state of the vehicles at any given instant.

This requirement emphasises the importance of gathering data that enriches the information obtained from the vehicles. Meeting this requirement not only implies the integration of additional data sources but also necessitates a sophisticated solution for storing the application data. In fact, vehicle data is typically presented as spatially contextualised time series – sequences of vehicle data points indexed in time order and enriched with GPS coordinates. This data is extremely dynamic and requires frequent acquisition, as the state of the vehicles constantly and rapidly changes. In contrast, the data that characterise vehicle-related entities is usually quasi-static, or at least slowly evolving, compared to vehicle measurements. This means that the system must manage two very different datasets, which require proper integration because the target applications rely on their combined use. Context-awareness therefore imposes a second quality constraint that directly affects the CVP structure:

(NFR2) The CVP shall employ a storage unit capable of handling both vehicle data and context data.

Context-aware, real-time fleet management is also subject to tight time constraints: fleet operations must be completed before their deadlines to avoid catastrophic consequences. To achieve this, the data platform must have access to always up-to-date and aligned data, enabling reactive processing whenever required. These time constraints therefore impose a non-functional requirement on the data acquisition process:

(NFR3) The CVP shall acquire and store the data required for processing as soon as it becomes available.

This translates into interventions at both the node and system level: at the node level, the structure of the data producers must be optimised as far as technologically possible using off-the-shelf hardware; at the system level, the specifics of the network used to transmit collected data to the remote repository must ensure timely and reliable delivery. However, ensuring responsiveness in data acquisition and efficiency of data procedures is not sufficient to keep the system's tasks on schedule, as the platform itself has a significant impact on data processing performance. Neglecting design considerations regarding the CVP – essential to understanding the real operating conditions of FMSs – is the primary error of several academic studies. These oversights, although unacceptable, are nevertheless totally understandable. Indeed, it is common practice to assume that the outstanding performance of existing processors is a silver bullet for the vast majority of ordinary applications. However, the overlapping of context-awareness requirements and real-time constraints creates performance challenges that lack a trivial technological solution. For example, the integration of datasets with diametrically opposed characteristics – such as vehicle and context data, which would normally require equally different data storage solutions – significantly hinders database performance. This is especially evident when performing complex queries, which are essential for context-aware, real-time applications. A final, though no less important, non-functional requirement can thus be defined as follows:

(NFR4) The CVP shall employ technologies capable of handling situation data efficiently.

The CVP typically employed for traditional fleet management has characteristics that differ from those discussed above, as it is based on a distinct set of non-functional requirements. As mentioned earlier, these applications rely solely on vehicle measurements, which determine the type of data producers required by the system – typically OBD-II dongles with a GPS interface. This, in turn, simplifies the design of the system's storage unit, since only vehicle data needs to be stored. Storage is usually handled via NoSQL databases, although relational databases are still occasionally used, particularly when complex queries are required, as discussed in Chapter 4. Moreover, no decision-making process needs to be completed under real-time constraints, meaning that no special considerations are required for either data acquisition or database performance. HTTP-based communication networks, which are not particularly reliable at low latency, can therefore be adopted, as detailed in Chapter 5.

Re-thinking the way we manage the decision-making process of certain fleets is of paramount importance, as such fleets have a profound impact on our lives. This requires an intervention in the structure of the FMSs. In fact, although all data platforms share a common structural layout, as they belong to the same industry vertical, they may still require different components and technologies. A visual comparison between the CVP used for traditional fleet management and that for context-aware, real-time applications is provided in Figure 2.5. This figure illustrates how, even when the system framework is practically identical, the structure of individual components, their interconnections, and the underlying technology can vary considerably depending on the non-functional requirements of the applications the FMSs must support.

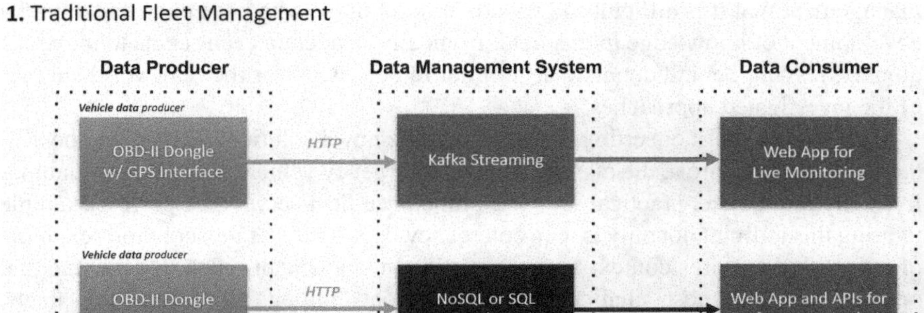

FIGURE 2.5 CVPs structural layout comparison.

2.2.4 Academic Approach to Emergency Fleet Management

Emergency management systems, such as the Emergency Medical Systems (EMS) employed in healthcare, are vital components of the complex ecosystem that is our society, as they have the delicate task of promptly responding to the onset of hazardous events, which can occur anywhere and at any time. Although the modus operandi of such systems varies across countries, they are invariably based on an emergency phone line that directs users' requests to a call centre. Each emergency request, often automatically classified according to risks, triggers a procedure for one or more vehicles to be dispatched. This operation is typically carried out by a radio operator, who sends the nearest available vehicles to their destination, normally with the support of AVL (Automatic Vehicle Locator) or CAD (Computer-Aided Dispatch) systems, which enable real-time monitoring of the fleet's location. Paradoxically, in an era of abundant information and advanced computing power, all vehicle operations in emergency services – from allocation to dispatch, as well as redeployment – still rely on rudimentary technologies and management rules. The academic literature on this topic, however, is far from stagnant; it is, in fact, rather fervent. So many studies have been published in the area of operational research for emergency vehicle management that it is difficult to give credit to all those who have contributed to the development of knowledge in this field. Ironically, academia even seems to move too quickly, leaving behind details that turn out to be crucial for the real-world success of the investigated approaches.

It is, however, disconcerting to realise that the novel solutions devised by the academic world to improve the management of emergency vehicle operations remain far from any immediate, practical use. The ambulance fleet serves as a perfect example to bring this to light: not only is it an emergency fleet – and the most obvious example of a fleet that requires context-aware, real-time management – but also the one that operates in the most challenging environment among all emergency fleets. Consequently, the insights gained from ambulance management often yield valuable lessons for other emergency services. Regarding ambulance allocation, one of the most noteworthy studies is by V. S. Knight et al., which proposes an algorithm capable of allocating emergency vehicles by optimising the overall expected survival probability of multiple classes of patients, rather than focusing only on average response time targets. Interestingly, the ability of this innovative strategy to meet real-world performance targets has been demonstrated in simulation only. A wide gap also persists between the theoretical models and practical implementation of various intelligent dispatching algorithms, largely due to the lack of opportunities to test these models in real operational environments. This issue is highlighted in the work of M. Amorim et al., which analyses, among others, the optimisation model for ambulance dispatch based on live traffic data developed by A. Haghani et al.: this model faces major challenges when applied in real situations due to its computational burden. On vehicle redeployment, worth mentioning is the method proposed by S. Enayati et al., which dynamically reallocates available ambulances after service completion, whose performance remains evaluated solely in simulation.

The above are mere examples of novel approaches to the management of emergency vehicle operations, which, however, underline the tendency of the academic research not to venture beyond the evaluation of the behaviour of algorithms per se. In other words, innovative methods are assessed in an unrealistic simulated environment, which omits elements that can influence their actual operating conditions. In addition, research topics are scattered and analysed individually. All this inevitably hampers not only the evaluation of the performance of the new strategies, but – above all – the possibility of adopting such solutions in our everyday life.

2.3 CHAPTER SUMMARY

This chapter has revealed the tendency of academic research to overlook the data management aspects required by fleet management applications. While this is generally not a major concern, such negligence becomes particularly significant in the case of context-aware, real-time fleet management solutions, as it hampers their practical implementation. Indeed, no connected data platform is currently capable of efficiently handling the unique data management required by these applications, which arises from the unusual overlap of context-awareness requirements and real-time constraints. Using the English NHS ambulance fleet as a reference, Chapter 3 reveals the pitfalls of flawed fleet data management and outlines the benefits that a bespoke strategy can bring to the performance of the services that depend on these vehicles.

3 Case Study
Ambulance Fleet

Ambulances play a decisive role in saving people's lives, as they serve as the primary means of rescue, first aid, and transport for the injured, sick, and wounded. In England alone, the publicly funded health system, often referred to as the NHS, responds to 10 million 999 calls every year [9], transporting around 40% of the patients admitted daily to English medical centres [10]. Responsiveness is clearly a central element for ambulances, as the speed of their response can make the difference between life and death. Such is the case with cardiac arrest, an emergency that requires patients to be reached in no more than seven minutes [11]. The tragic event of the terrorist attack at the Manchester Arena saw the first ambulance on the scene in the minuscule time of one minute after the explosion of the bomb. Although it is natural to be pleasantly impressed by the invaluable contribution of ambulances, it is impossible not to notice a slow but steady decline in the performance of the services that rely on this vehicle fleet. Regrettably, the demand for ambulance services continues to grow rapidly, with an average annual increase of around 5.2% [12]. For this reason, an ever-increasing amount of funds is being allocated to reinforce the emergency service, which is still not sufficient to meet this growing demand. This became glaringly obvious during the COVID-19 outbreak: in the winter of 2021/22, with the surge of the Omicron variant, the NHS saw unprecedented demand for its services and immense difficulties in maintaining the patient flow. In January 2022, there were over 15,000 delays of more than two hours, which is almost four times the delays for the same period in 2021 [13]. Ambulance services have been at the forefront of technological innovation for many years, but the pressure they are now under inevitably raises questions about their model. It is therefore time to revise the way ambulance services are handled, and perhaps the right starting point would be the data platform that enables their decision-making.

The main objective of this chapter is to illustrate the flaws of the current approach to ambulance management and the benefits that a tailor-made data-driven strategy can bring. While the section "Service Operation and Current Performance" focuses on the present operational productivity of the English NHS ambulance service, the section "Designing a CVP for Ambulance Fleet Management" takes the first steps in the design of the data platform required to achieve a better-performing ambulance management.

3.1 SERVICE OPERATION AND CURRENT PERFORMANCE

The demand for ambulance services has dramatically increased over the last few years and is likely to continue to soar. To adequately meet this growing demand, rather than systematically increasing budgets and manpower, ambulance services

DOI: 10.1201/9781003688549-3

should be fine-tuned to generate more output with the same amount of resources. Addressing their operational productivity would in fact ensure a reduction in the unsustainable costs they are incurring and, above all, a better outcome for patients. The purpose of this section is to examine how the NHS is presently managing its ambulance service, with the aim of identifying areas where action can be taken to improve its performance. Obviously, the focus is on the performance issues that a poor fleet data management brings with it. This inevitably leads to the exclusion of all those factors that, although related to the performance of the ambulance service, are not directly attributable to vehicle management, like staff wellbeing. Hence, after a concise overview of the present performance of the English NHS ambulance service and the effect that data has on it, two subsections will follow, echoing the classification of fleet management applications proposed in Chapter 1: the first subsection describes the NHS strategies around performance assessment, while the second covers the way monitoring and vehicle operations like dispatch are orchestrated, which are discussed together as they are enabled by the same technology.

3.1.1 The Importance of Data for Ambulance Operations

The NHS Ambulance Service provides unplanned assistance to nearly 67 million citizens across the UK, 24 hours a day, 365 days a year. As shown in Figure 3.1, ten regional ambulance trusts are currently active in England, with a separate organisation for the Isle of Wight – plus three ambulance services running in Wales, Scotland, and Northern Ireland. To date, there is no common operating procedure among British trusts, which leads to considerable differences in their performance. However, while it is extremely difficult to compare NHS trust performance – since no method exists to account for demographic and geographic variations in the area they serve – it is beyond any doubt that their overall effectiveness has declined. Surprisingly, this decline has never been linked to poor data management.

FIGURE 3.1 Ambulance service areas in the UK [14].

The present section explores the management of English NHS ambulance fleets only.

Since 1975, the performance of the ambulance service has been gauged against a set of time-based targets, based on the understanding that patients with potentially life-threatening emergencies have a higher chance of survival if they receive a faster response. However, in recent years, this rather simplistic approach – combined with a dramatic but unremitting increase in the volume of requests – has negatively affected the performance of the ambulance service. Until 2017, NHS ambulances were held to an eight-minute response time target for virtually any call, even though not all emergencies required immediate action. The fallout was to put call handlers under undue pressure and to prompt NHS trusts to activate multiple ambulances – often unnecessarily – in an attempt to meet such a stringent target. Noting that life-threatening calls account for no more than 10% of the total, the NHS agreed on a new performance standard for its ambulance service, establishing different response time targets according to the urgency of each call:

- *7 minutes mean response time* for **life-threatening calls**, which are time-critical events requiring immediate intervention, such as cardiac arrests.
- *18 minutes mean response time* for **emergency calls**, which are potentially serious conditions that may require rapid assessment and urgent on-scene intervention and transport, such as strokes and major burns.
- *120 minutes maximum response time* for **urgent calls**, which are problems that require treatment to relieve suffering, as well as transport within a clinically appropriate time frame, such as falls and diabetic problems.
- *180 minutes maximum response time* for **less urgent calls**, which are conditions that are less compelling but may still require assessment and possibly transport within an appropriate time frame, such as diarrhoea.

The new approach gives now call handlers additional time to spend on each call, with the exception of life-threatening emergencies, which still require the fastest response. Research by the University of Sheffield concluded that this model could save 250 extra lives per year and provide a substantial boost in the performance of the ambulance service, which evolved from focusing solely on transferring patients to the hospital to ensuring that patients receive the right care for their needs. Despite the introduction of the standard, however, performance remains mixed. Evidence of this can be seen in the inability of eight English NHS trusts, as of July 2018, to meet the response targets for life-threatening and emergency calls, as illustrated in Figure 3.2 – not to mention the performance challenges experienced by the ambulance service during the recent coronavirus pandemic. Alarmingly, these figures suggest that, although the introduction of a new ambulance performance assessment standard was desperately needed, the causes behind the decline in the ambulance service efficiency must lie elsewhere.

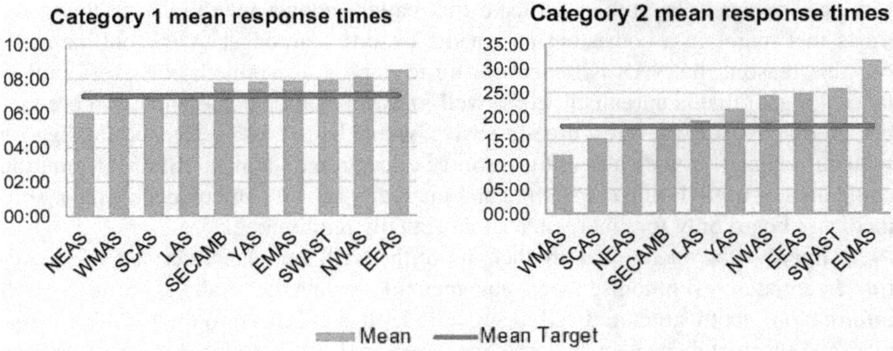

FIGURE 3.2 Ambulance service response time in minutes, July 2018 [10].

Albeit in part, this decline can undeniably be traced to flawed data management. In fact, the operational productivity of the ambulance service can be drastically improved through a more conscious and structured use of available information. Evidence of this emerges from a multitude of initiatives, such as the Data Awareness for Sending Help project, which was launched in 2017 through a collaboration between the Computer Science Department at King's College and the London Ambulance Service. The project aimed to establish policy recommendations for the use of new and emerging data sources to improve the ambulance service performance. It identified a variety of data sets that could enhance the capabilities of dispatch decision-support systems, thereby actively contributing to faster ambulance response times. The initiative highlights the potential to capitalise on the usually overwhelming *data deluge*, exploiting real-time insights from multiple sources to reinforce the data foundation on which services are built. According to the study, the data already in use by commercial CAD systems – such as vehicle status and location – can be combined with transport systems data, air quality monitoring, weather forecasts, and mobile phone location, to enable a more effective decision-making for ambulance dispatch.

Initiatives such as the one mentioned above are clearly desirable, as they contribute to improving the performance of the ambulance service. However, given the current management of ambulance service data, such approaches could create the paradoxical situation in which external data is fully exploited, while much of the information that is already available to the NHS remains untapped. Yet, this situation should not be attributed to malpractice by the NHS, but rather to a genuine inability to fully understand – and therefore properly manage – the high data complexity that arises from the knotty environment in which ambulances operate. For example, the NHS insists on separating context data, such as staff information, from vehicle data, as it is not considered directly relevant to the on-road operations of ambulances. As a result, no mechanism exists for the combined management of vehicle and context data, with the latter usually scattered across several data units without any standard procedure or, worse still, on infrequently updated paper documents. This inability to correlate the two datasets inevitably impairs the performance of the ambulance service. For example, studies indicate that patient conveyance rates are lower for trusts that maintain a consistent paramedic expertise amongst their frontline staff. For this reason, the NHS is endeavouring to deploy a paramedic on every ambulance – a goal that is unrealistic given well-known resource constraints. A more viable alternative, since less resource-intensive, would be to cross-reference call history with current staff rota data to tailor resource allocation to demand. Staff information could also be used during dispatch to automatically activate ambulances with a paramedic on board only for emergencies that actually require one.

To improve the operational productivity of the NHS ambulance service, it is therefore essential to revolutionise the management of the data that is already available. All information – both structured and unstructured – that is related to the ambulance service, and specifically its vehicle fleet, should be seamlessly integrated. Ambulance data could then be considered *big data*, whose use can enable functionalities capable of skyrocketing service performance. As detailed in the following subsections, all fleet management activities can reap significant benefits from the combined use of vehicle and context data. For example, performance evaluation could leverage predictive analysis, cluster identification, and knowledge discovery techniques; access to exhaustive fleet information and the ability to process it efficiently would enable automated decision-making, which could provide timely alerts during monitoring activities and fully automate crucial operations like dispatch. Hence, the need for a brand-new CVP.

3.1.2 Strategies to Assess Fleet Performance

The English NHS ambulance fleet entails an annual expense of approximately £200 million, excluding the annual capital investment in new vehicles, making it the largest area of non-pay expenditure across the entire ambulance service. It is therefore not surprising that the NHS explicitly encourages trusts to undertake performance assessment activities to optimise resources, reduce spare parts, and avoid ambulance breakdowns. Such measures could result in significant cost savings and, above all, in a more resilient service for patients. Yet, despite this, there is no consistent practice in the use of IT and data management systems, with fewer than half of the trusts currently employing a traditional fleet management solution. At the present time, it is thus inconceivable to even contemplate the adoption of sophisticated fleet performance analyses, such as those based on the combined use of vehicle and context data.

The NHS operates a fleet of over 5000 vehicles, approximately divided between 3200 Double Crewed Ambulances (DCA), 1500 Rapid Response Vehicles (RVV), and a small number of other classes of vehicles like motorbikes and specialised units. Each vehicle is estimated to travel around 50,000 miles per year, on average, five times the annual mileage of a personal car. Moreover, since ambulances regularly operate under far more demanding driving conditions than passenger vehicles, their lifespan is relatively limited. Premature vehicle ageing is a major cost concern for the ambulance service, with approximately 2500 units replaced every five years. Delaying vehicle replacement is, however, utterly detrimental, as an older fleet not only incurs higher maintenance costs – ranging from £4,200 to £6,900 per vehicle per year – but also reduces service reliability, given the increased likelihood of malfunctions.

An effective strategy for reducing fleet expenses without compromising performance is to improve vehicle maintenance through traditional fleet management services. By constantly monitoring vehicles' status and issuing responsive alerts when malfunctions occur, these services can potentially reduce fleet maintenance costs by 20%. They also deliver significant benefits to fuel management: by providing insights on driving behaviour and recommending the replacement of faulty vehicles, they can achieve fuel savings of up to 14% on DCAs and 22% on RRVs. Surprisingly, fewer than half of the ambulances currently record vehicle data, and even when this information is collected, it is not always monitored or used effectively.

As already noted, traditional fleet management services are designed for a specific class of vehicles, which – unlike ambulances – do not require context data beyond GPS coordinates. Consequently, even if all trusts were to adopt such products, much of the data related to ambulance operations would still inevitably be lost, or at best, not easily accessible. For instance, even when using fleet management services, NHS trusts do not consistently record the amount of time vehicles are off-road, nor the reason behind it, whether due to mechanical failures or accidents. This is illogical, given the opportunities that context data could unlock: correlating vehicle diagnostics data with maintenance history could help anticipate breakdowns, while cross-referencing driving behaviour with call categories data could lead to accurate forecasts of the spare parts required for each vehicle class. There would also be benefits for activities not directly related to on-road operations: the combined analysis of call patterns and the local factors influencing them – such as demographics, seasonality, weather, and even tourism – could provide insights into the optimal staff rota to meet call forecasts and their seasonal variations. The purpose of these activities is thus to guide resource allocation through effective modelling, which evidently requires context data too.

3.1.3 Fleet Monitoring and Operations Management

Control centres form the backbone of the ambulance service: they act as the gateway to medical support, as they are responsible for prioritising calls, giving first aid advice over the phone, and coordinating frontline responses to emergencies. With over 10 million calls to answer every year, these centres require a staggering number of operators, who must make delicate decisions at an extremely fast pace while dealing with individuals in distress. It is definitely a challenging environment, made even more demanding by the sharp increase in the volume of emergency calls: staff often struggle to cope with this overwhelming burden, which inevitably leads to a significant deterioration in the performance of the ambulance service. According to the NHS, the only possible solution is to expand the control centres workforce – an approach that clearly conflicts with the *more-for-the-same* direction imposed by limited financial resources. A more reasonable approach would be to completely rearrange the data management behind control centres' operations, a shift could drastically lighten staff workload. Yet, the NHS seems to be blind to the prospect of any technological shift.

Across its 22 centres, the English NHS presently employs a workforce of 3800 call handlers, call dispatchers, and clinicians, each serving a particular function. The typical activity of an ambulance control centre begins with the reception of a call, followed by the collection of all information necessary to accurately establish the severity of the emergency. This operation, formally known as call triage, is carried out by the call handler with the support of an Advanced Medical Priority Dispatch System (AMPDS), which prioritises the intervention based on the gathered data and supports the operator in choosing the appropriate response. Depending on the severity of the case, the call is then directed either to a clinician, who provides advice and refers to a suitable alternative service, or to a dispatcher, who assigns the most appropriate mobile unit according to the immediate availability. Like call handlers, dispatchers rely on a sophisticated software platform, generally referred to as CAD. This complex software suite allows operators to monitor the available units via an AVL module, notify the appropriate resources through a dedicated radio channel, and maintain the status of responding vehicles in the field. The choice of the vehicle to be dispatched is assisted by an additional software module, which provides the operator with a list of the available vehicles sorted by their distance from the incident. In fact, because the available data is far from complete, dispatch cannot be fully automated, but is left to the ability of the operator. This inevitably exposes the decision-making process to errors. This is the case, for example, with calls requiring a swift response that leaves no room for further assessment: if no information is available on the ambulance crew and their skill level, the operator may inadvertently assign a unit that is unsuitable to serve the emergency. When the only data provided is the vehicle's position, the operator has no alternative but to send the nearest available ambulance, which may not always represent the optimal solution. Once a vehicle has been selected, the call handler notifies the relevant resources and remains in communication with them for the duration of the call.

Despite being one of the pillars of the ambulance service, control centres remain one of its greatest weaknesses: their imperfect decision-making, in fact, has tremendous repercussions on the operational productivity of the overall service. A major contributor to this deficiency is poor data management. As discussed earlier, the absence of an exhaustive description of the state of the ambulance fleet is a key barrier to the adoption of automatic decision-makers. Because current decision-making is human-based and therefore not systematic, control centres inevitably dispatch more vehicles than necessary. For example, there is a notable difference in the number of units activated by the different ambulance trusts: in 2017, for every 100 incidents, trusts sent an average of 137–174 vehicles, a variation so significant that it cannot be superficially attributed to a difference in call profiles. Automatic decision-makers would not only eliminate unnecessary ambulance activations, but also significantly reduce the Job Cycle Time (JCT) – the total time taken to respond to an incident – since the dispatch procedure would also become virtually instantaneous.

More importantly, automatic decision-makers would relieve call dispatchers of an increasingly unbearable strain as demand expands unabated. It is estimated that by 2028, the English NHS ambulance service could be handling 4 million additional calls, which equates to an additional 1500 control staff under the current configuration. Given that the existing workforce presently costs nearly £50 million per annum and is considered underpaid and difficult to retain, such a strategy appears impractical. A tailor-made data management, on the other hand, could boost control centres' performance, freeing human operators from decisions that are invariably prone to errors and delays, and enabling automated, near-instantaneous, and error-free decision-making.

The importance of data is being acknowledged by the NHS itself, which is paving the way for the collection and analysis of a broader range of ambulance dispatch information, consistent across the ambulance trusts nationwide. However, this must be accompanied by an adaptation of the underlying CVP, so that trusts can not only collect both vehicle and context data but also manage them in a timely manner. This is crucial, as the delicate decision-making at the heart of ambulance service operations is perilously time-critical.

3.2 DESIGNING A CVP FOR AMBULANCE FLEET MANAGEMENT

The performance of the ambulance service continues to deteriorate alarmingly as the volume of calls rises. Given the vital role of ambulances in maintaining public health, reversing this trend is a compulsory requirement. This can be achieved by reshaping the data platform on which the ambulance service is built, since its performance is profoundly affected by the way the fleet data is managed. However, reframing the ambulance service's data management is far from simple, as none of the existing data platforms can adequately support context-aware, real-time fleet management applications. The reason for this lies in the unconventional data management required by this particular type of fleet, which inevitably influences the structure of the underlying data platform. In fact, although all CVPs share a common structural outline, each data platform must be modelled around a different set of non-functional requirements – namely, the quality constraints that must be satisfied for the management of the application to be effective. The ambulance service, in particular, requires the efficient management of both vehicle and context data, which is not required by any mainstream fleet management application. It is therefore necessary to build on these requirements and develop a tailor-made data platform that can improve the operational productivity of the ambulance service. This means enhancing the performance of the service without expanding its workforce, an unfeasible option given the limited financial resources available.

The purpose of this section is to outline the design of a bespoke CVP that goes beyond the operational capabilities of existing data platforms. The goal is to improve the decision-making processes that underpin the various operations of the NHS ambulance service. For the development of the data platform, we have adopted an *iterative and incremental approach*, which consists of implementing small portions of end-to-end functionality through repeated cycles. This approach is preferable to more rigid methodologies such as the waterfall model, which requires extensive planning. In a project like the one presented in this book, where the system requirements are not entirely clear from the outset, such planning is indeed impractical. By contrast, the chosen methodology, based on shorter development cycles, allows for faster feedback, facilitating the clarification of vague requirements during the course of the project. The possibility of incorporating amendments at the end of each iteration also enables a faster and more efficient response to change. With this approach, each development cycle results in the implementation, alteration, or refinement of a prototype, which is an incomplete version of one of the components of the data platform. Although it is inevitable that some prototypes will be discarded, this iterative process ultimately establishes a solid groundwork for the definitive version of the CVP.

By following the typical structure of a Software Requirement Specification (SRS) document, this section discusses the purpose of the system, its functionalities, the data required for their implementation, the intended users, and the interfaces through which these users can access the system's functionalities. It is important to note that, due to the chosen software development model, the documentation that follows is not definitive and may be subject to future expansions and modifications, as the CVP development process may undergo further iterations.

3.2.1 Purpose and Intended Use

The CVP is intended for the management of the ambulance fleet of the English NHS. Accordingly, the data platform must be capable of transforming any information related to the ambulance service into informed decisions regarding the fleet. The system will therefore support the acquisition of *vehicle data*, as provided by traditional FMS. It will also acquire *context data* – information that characterises the situations in which these vehicles operate, which is typically neglected by existing commercial data platforms. The combined analysis of these two datasets, collectively referred to as situation data, will assist in any decision concerning the vehicles' life cycles. End-users must therefore be able to *assess* ambulance service performance, *monitor* collected data, and *manage* vehicle operations such as dispatch. A black-box representation of the CVP, which focuses on the functions offered to the end-users and the data required to enable them, is provided in Figure 3.3.

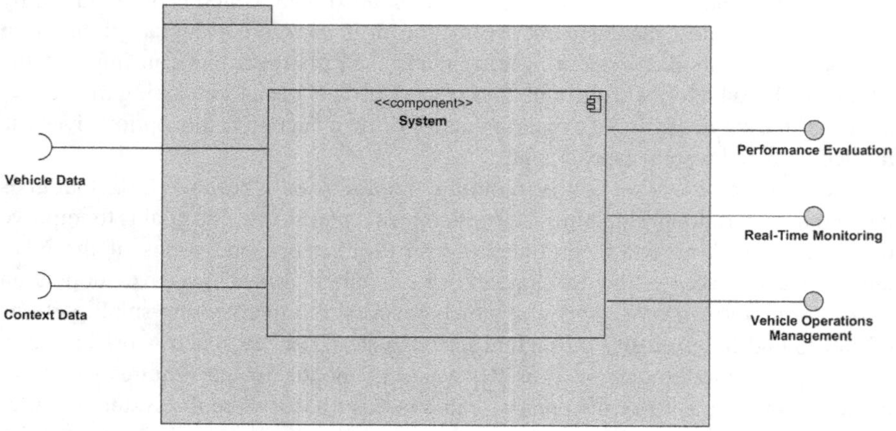

FIGURE 3.3 CVP: Black-box representation.

3.2.2 System Features

Before embarking on system development, it is desirable to clarify the functionalities that the final product will provide. As per good software engineering practice, this can be achieved through functional requirements, which describe a system function as a specification of behaviour between inputs and outputs. Functional requirements thus provide a usable representation of a need, as they define the features that the system must necessarily support to ensure the end-users can achieve their goals through its use. These requirements lay the foundation for the development of the final product, as they enable developers to delineate the system domain, the users of its functions, and the data required for their implementation. However, it should be emphasised that functional requirements are always specific to the particular application the system is intended to support. Unlike non-functional requirements, they do not characterise any form of context-aware, real-time fleet management, but are contingent on the chosen application. Although they may seem of limited relevance to the scope of this project due to their lack of generality, functional requirements are indispensable: without them, it would be impossible to implement a prototype of the data platform, upon which compliance with the non-functional requirements can subsequently be assessed.

The aim of this subsection is to determine the functional requirements of a CVP for the effective management of the ambulance fleet. This represents a further step towards system development: while domain requirements lead to the fundamental structure of the data platform, functional requirements define the interfaces for data collection, the logic for data processing, and the terminals for end-users. This then allows the design and development of the individual components of the CVP: at this stage, it is essential to ensure that the system provides the intended functionalities while complying with the non-functional requirements outlined in Chapter 2.

According to the definition of fleet management presented in Section 2.1.1, the CVP must be capable of supporting any decisions concerning the fleet it is intended for. As in the traditional fleet management approach, the data platform will provide fleet managers with the ability to *assess vehicle performance*. However, because the operating environment of ambulances is more complex than that of commercial vehicles, the CVP must ensure that the behaviour of the units can be correlated with the data of the various entities related to them. Yet, unlike traditional commercial solutions, ambulance fleet management cannot be limited to performance evaluation activities alone. As discussed in Section 3.1, a significant portion of ambulance service decision-making takes place in the control centres. This means that the CVP must also support control-centre operators in all decisions concerning vehicle operations during emergency situations. Specifically, it must allow users to *monitor in real-time* the status of the units, their operations, and the staff assigned to them, as well as supervise the *automated dispatch* of vehicles in response to the reception of a call. The dispatch procedure will be initiated by call handlers, who provide the system with information about the patient, the intervention to be managed, and the severity of the emergency – that is, the outcome of the call triage. The system will then assign the call to the quickest available vehicle, which must be operated by a crew qualified to handle the emergency, and inform the staff operating it via a dedicated communication channel.

To deliver all of the above functions, the system requires the combined use of vehicle and staff data. In addition to a set of dongles interfacing with the vehicles, the CVP should include a component that allows ambulance operators to provide their professional data and skill level, and enables the system to keep track of their ambulance shifts. This component will also allow operators to receive dispatch notifications and to close an active call once the emergency has been resolved. Ambulance operators should, in fact, be able to inform the system of the new availability of a vehicle regardless of its location – a feature that is particularly in view of the recent introduction of new ambulance care models. Although most emergency calls result in the activation of an ambulance, not all patients are in fact transported to the Accident and Emergency (A&E) department, as required by the conventional *See&Convey* model. Approximately half of the patients are treated and discharged on scene under the new *See&Treat* response model, meaning a vehicle can become available before returning to its hospital.

Hence, the CVP will provide the usual fleet management activities, adapted to the ambulance context, plus a number of support functions, as summarised below:

- *Performance assessment*: the system must allow fleet managers to evaluate the ambulance service's performance through the analysis of situation data.
- *Fleet status live monitoring*: the system must allow fleet managers and call handlers to obtain a real-time overview of the fleet's status, which consists of both vehicle data and context data, by specifying the area of interest.
- *Automated vehicle dispatch*: the system must allow call handlers to initiate the automated dispatch procedure by providing details of the emergency. This involves sending a notification to the staff operating the dispatched vehicle.
- *Shift management*: the system must allow ambulance operators to start – and then eventually stop – an ambulance shift by providing details of the vehicle they will be operating. Both functions require ambulance operators to have created a personal profile and logged into the system.
- *Call Management*: the system must enable ambulance operators to conclude the call they are serving once the emergency has been resolved, and allow them to be promptly notified of any changes in the operational status of their unit. Both features require vehicle operators to have started an ambulance shift.

The above-mentioned functions can also be depicted using one of the many graphical representations offered by the Unified Modelling Language (UML), which provides designers with standard diagrams to simplify the specification, visualisation, construction, and documentation of a system. Figure 3.4 presents a use case diagram illustrating the functions of the data platform, along with the actors and external systems that interact with it. The details of these elements are discussed in the following subsections.

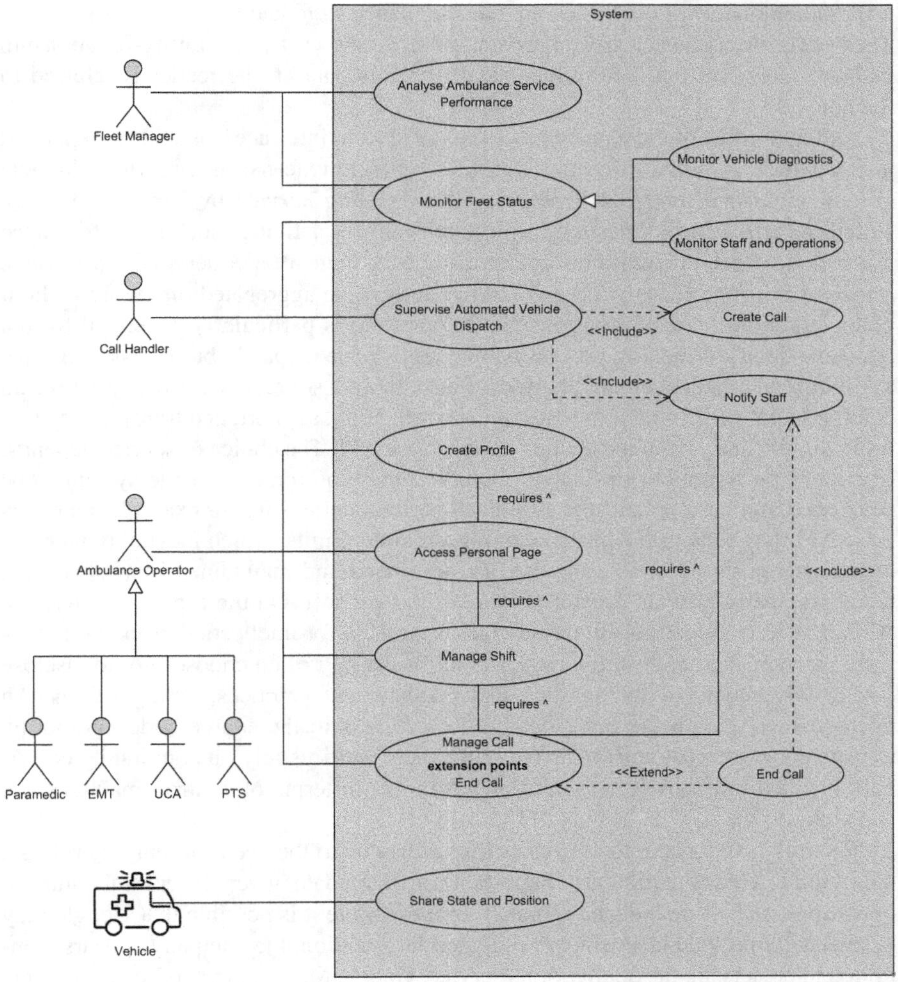

FIGURE 3.4 CVP: Use case diagram.

3.2.3 System Data

Data represents the smallest units of factual information that can be used as a basis for calculations, discussions, and reasoning. It plays a central role in the functioning of every system, acting as the raw material that the system receives, analyses, and elaborates to generate meaningful outputs. As discussed at the outset of this work, any context-aware, real-time fleet management application requires two different datasets: vehicle data and context data. The former, which consists of the *measurements* that describe the status of a fleet vehicle at a given moment, is common to all fleet management applications. The latter, by contrast, varies, as context changes with the application. To enable the functionalities outlined in the previous subsection, contextual information on the *vehicles* and the *dongles* connected to them, as well as information on the *call* to be served, the ambulance *staff*, and their *shifts*, is required. These data items are clearly interconnected, as shown by the Entity-Relationship diagram in Figure 3.5. For further details on this dataset, the reader is referred to Section 4.2.

Although most of the data comes from the system interfaces, as will be explained in Section 3.2.5, some information is sourced from external data services. Vehicle specification data, obtained from the *VIN decoding service* by Vincario, will be stored in the platform repository, as it is relevant for activities such as performance evaluation, which processes information at a later stage after collection. Traffic data, retrieved from the *Google Maps APIs*, can instead be aggregated on the fly without being stored in the data platform; this information is particularly beneficial for the real-time decision-making process behind ambulance dispatch, but is less useful for traditional fleet management analyses. For both data sources, we have opted for an enterprise API, with characteristics of cost, reliability, support, and transparency that differ significantly from those of an open-source API. The choice of service depends, of course, on a number of factors, including the requirements of the system to be developed and the needs of the organisation that operates it. For example, an enterprise API may require payments and impose usage limits, which inevitably increase operational costs. However, these APIs, developed and maintained by professional teams, typically provide greater reliability and support. On the flip side, enterprise APIs may have limitations in terms of transparency: for intellectual property or strategic reasons, the organisation responsible for the API often chooses not to disclose any details about its functioning, data management practices, or algorithms. An open-source API, on the other hand, allows access to the source code, promoting transparency and customisation. Yet, since these services rely on community contributions, they may have significant shortcomings in terms of control, maintenance, and support.

Naturally, we have opted for enterprise APIs due to the need for a highly reliable data source. This is particularly important for traffic data, given the critical nature of operations such as ambulance dispatch. Indeed, while it is possible that Google may occasionally provide inaccurate or outdated information, the company appears committed to preserving the quality of its service. Google Maps, in fact, relies on its users to report road closures, who may maliciously alter the data provided by the APIs. User reports, however, are reviewed by both automated systems and human moderators. The company also makes use of historical data and machine learning algorithms to identify patterns and anomalies in these reports, allowing it to detect and flag potential false information. In addition, Google aggregates data from reliable sources, such as government agencies, official road authorities, and verified data providers, whose cross-checking helps validate data accuracy. Therefore, choosing this service over an open-source API seems highly justified.

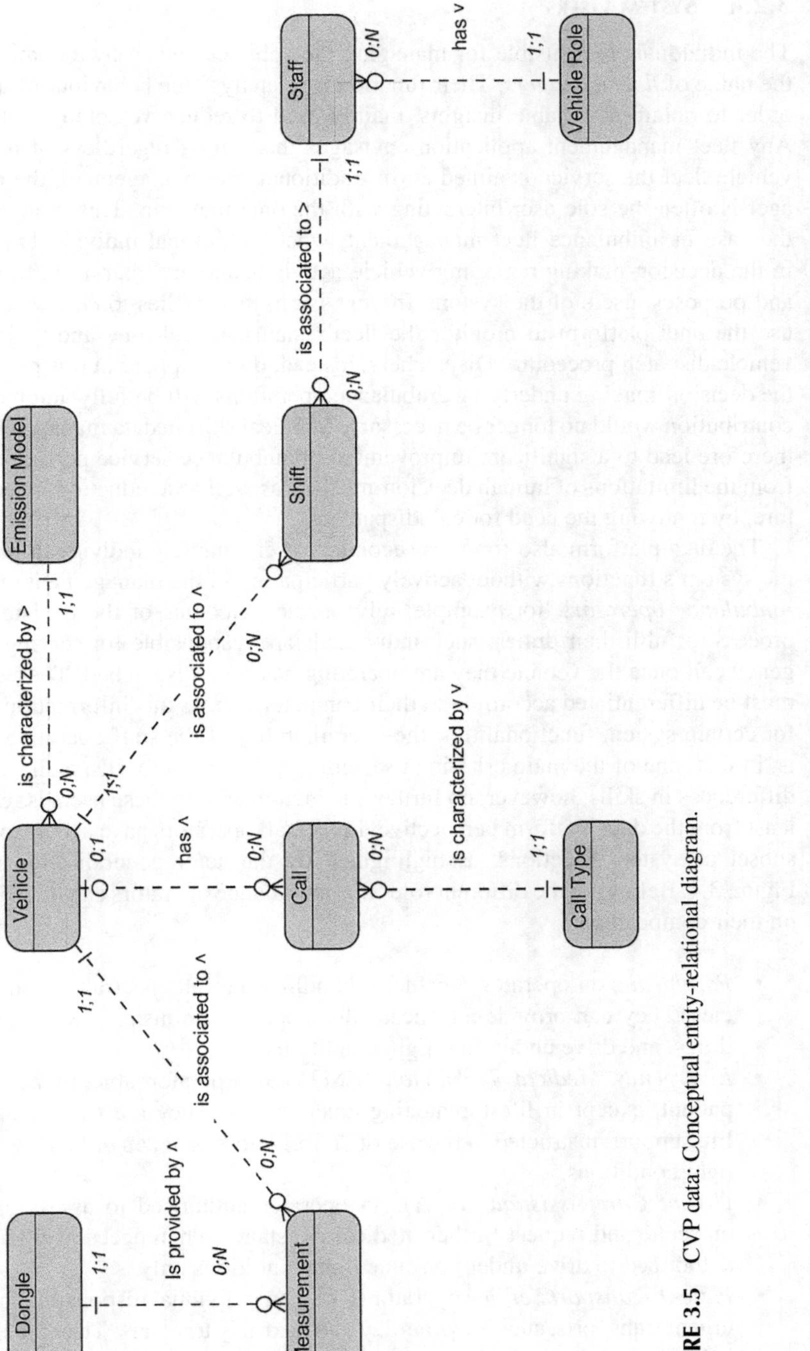

FIGURE 3.5 CVP data: Conceptual entity-relational diagram.

3.2.4 System Users

The individuals responsible for managing the vehicles owned by a company go by the name of *fleet managers*. Their function is to analyse the behaviour of the fleet in order to obtain actionable insights, mainly used to reduce vehicle operating costs. Any fleet management application envisages this figure, regardless of the specific vehicle fleet the service is aimed at. In traditional fleet management, the fleet manager is often the sole user interacting with the data platform. This, however, is not the case in ambulance fleet management, where additional individuals participate in the decision-making regarding vehicle activities and are, therefore, to all intents and purposes, users of the system. In our vision, this applies to *call handlers*, who use the data platform to monitor the fleet's status in real-time and to initiate the vehicle dispatch procedure. Dispatchers, instead, do not appear in this picture: since the decision-making underlying ambulance operations will be fully automated, their contribution would no longer be necessary. A radical shift in data management would therefore lead to a significant improvement in ambulance service performance, free from the limitations of human decision-making, as well as a reduction in its expenditure, by removing the need for call dispatchers.

The data platform also foresees secondary users, namely individuals who utilise the system's functions without actively participating in the management of the fleet. *Ambulance operators*, for example, rely on the outcomes of the decision-making process to fulfil their duties: such individuals are responsible for serving an emergency call once the vehicle they are operating has been dispatched. These operators must be differentiated according to their competencies, as this information is pivotal for certain system functionalities: the overall ability of the staff operating a vehicle is, in fact, one of the main criteria in selecting ambulances for dispatch. Apart from differences in skills, however, no further distinction among these users is required, at least from the data platform perspective. In fact, all operators have access to the same subset of system functions, as highlighted by the actor generalisation shown in Figure 3.4. Below are the different roles that ambulance operators can hold depending on their competencies:

- *Paramedic*: an operator capable of handling the full spectrum of emergencies. They can provide advanced life support, administer a wide range of drugs, and drive under blue-light conditions.
- *Emergency Medical Technician* (EMT): an operator able to assist any patient, except in life-threatening emergencies. They are trained in basic life support, instructed in the use of defibrillators, and can drive under blue-light conditions.
- *Urgent Care Assistant* (UCA): an operator authorised to assess patients on scene and request further medical assistance when necessary. They are authorised to drive under non-blue light conditions only.
- *Patient Transport Service* operator (PTS): an individual responsible for non-urgent transports, such as same-day and next-day transfers. They are trained in basic life support and can only drive under non-blue light conditions.

Fleet vehicles, on the other hand, do not qualify as end-users of the system: although they interact with the platform, they do not resort to its functions. Ambulances should therefore be regarded as external systems, as they are mere data sources for the CVP. Nonetheless, they still require a system interface just like any data platform user, the features of which will be discussed in the next section.

3.2.5 System Interfaces

Any exchange of information between the data platform and the various entities that interact with it is enabled by interfaces, which can be seen as a shared boundary between the system and the outside world. Since the numerous entities that interact with the data platform – whether human users or other data processing systems - do so in different ways, each of them inevitably requires a distinct interface. Fleet managers, for example, interface with the system to obtain information on the performance of the ambulance service and to monitor the status of its fleet. It is then necessary to provide an input/output interface that allows such users to submit a request to the data platform and obtain the corresponding result. This interface can be implemented as a *web application* – that is, an application program stored on a remote server and accessible via the Internet through a web browser. Such a system interface would allow multiple fleet managers to access the same version of the service on different platforms, including desktops, laptops, or mobile phones, without requiring any installation.

It goes without saying that the system also foresees a *web application* for call handlers, which allows them to provide details of emergencies and initiate the automatic dispatch of ambulances. This web app is also responsible for connecting the data platform with external services, such as Google Maps: it requests live data on traffic conditions, roadworks, and potential incidents along the route to help the system identify the vehicle with the shortest travel time to a given destination. Since both fleet managers and call handlers require access to the fleet status monitoring service, an equally plausible solution would be to implement a single web application that provides access to different system functionalities based on the user's role, following a preliminary identification and authentication process.

The interface used by ambulance operators will inevitably differ from that provided to fleet managers and call handlers. However, although the interaction of these users with the system is fundamentally different, their interface could still be implemented as a web application. This would allow system functions to be accessed directly from the operators' mobile devices, regardless of the underlying operating system. In addition to its broad compatibility, a web application offers other benefits, such as the ability to easily update system functionalities, since a mobile website provides a high degree of flexibility in content updates. Nonetheless, despite the many advantages of the mobile web, certain use scenarios make the adoption of a native mobile application preferable. This is despite the substantial challenges that the development of a mobile application brings with it, such as higher development costs, since each mobile platform – like iOS and Android – requires a separate software application. Nevertheless, a mobile app is particularly advantageous for our system, as it enables access to the native components of the user's device, such as the camera and GPS. It also allows better management of the phone's processing power and more efficient use of its battery, which is crucial considering that the application would run throughout the entire shift of ambulance operators. Another inherent capability of the mobile application is its ability to deliver push notifications even when the user interface is not actively displayed on the screen. This feature, essential for our data platform, would be extremely difficult to implement in a web application. Ambulance operators will therefore be equipped with a *mobile app* to interact with the system. This interface functions both as a data producer – sharing staff information with the CVP – and as a data consumer – providing ambulance operators with the result of dispatch operations as well as notifications of any changes in the operational status of their vehicle.

As previously stated, vehicles are not users but external systems to the data platform; however, they still require a system interface to share their information with the CVP. The vehicle interface should, of course, allow the collection and transmission of all data needed by fleet managers to analyse fleet performance. This includes ambulance diagnostics data and key vehicle attributes such as weight, engine, and fuel type. Much of this information can be retrieved via the standard OBD-II connector, which vehicles are equipped with, along with one of the diagnostic protocols this connector supports. The connector, in fact, provides access to the ambulance's on-board network, responsible for the exchange of data between the vehicle's control units. However, the vehicle interface should also communicate with external services to obtain data that cannot be acquired from the in-vehicle network. For instance, a VIN decoding API allows a wide variety of information to be derived from the vehicle identification number (VIN), which can be obtained together with diagnostic data. The location of the ambulance must also be captured, as this information plays a critical role in both monitoring and dispatching operations. To this end, our vehicle interface – implemented in the form of an OBD-II dongle – will be complemented with a tracker for real-time GPS position tagging, which has proven sufficiently effective for the target application.

The interfaces offered by the data platform, along with the users and external systems that interact with them, are shown in Figure 3.6. A new terminology is also introduced, as the analysis of the system interfaces has revealed that the previous classification, which distinguishes between data producers and consumers, is no longer sufficient. Indeed, the interface used by the ambulance operators supports both data collection and the presentation of the results of the system's processing. Consequently, a revised terminology is adopted: the dongle and the mobile application are grouped under the label **vehicle level**, as both interfaces are functionally related to ambulance operations, while the two web applications are classified under **service level**, as they are responsible for delivering the main functions of the data platform.

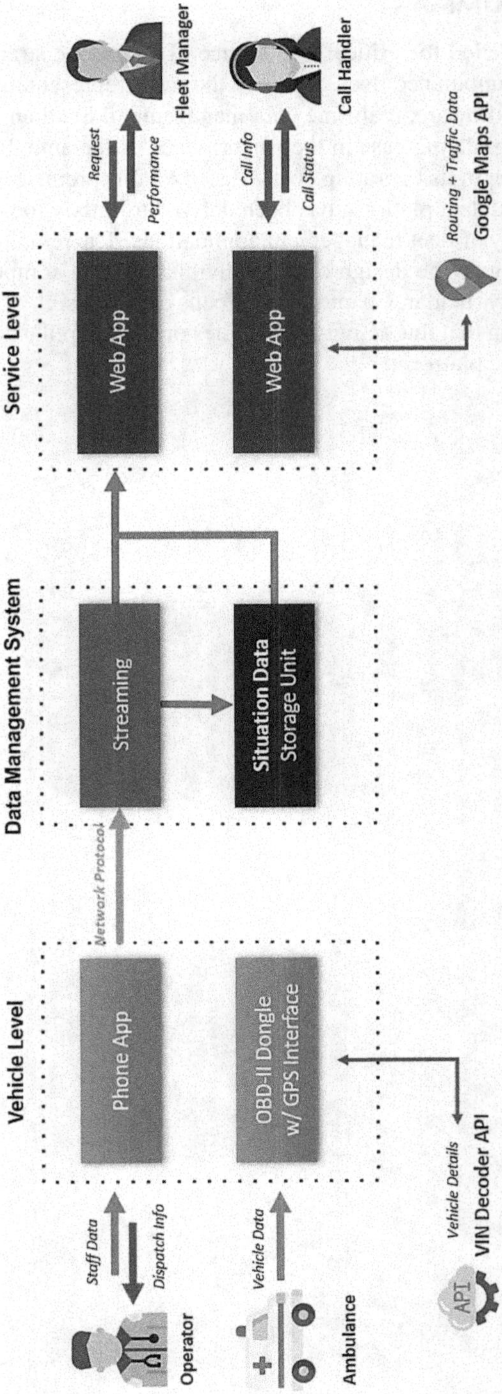

FIGURE 3.6 CVP architecture: Users and interfaces.

3.3 CHAPTER SUMMARY

This chapter has unveiled the critical importance of a bespoke strategy for the data management of the ambulance fleet, arguably the most representative example of a fleet requiring context-aware, real-time data management. Such an approach would guarantee both a tangible increase in the performance of the ambulance service and a significant reduction in its operating costs. Based on this premise, the fundamental design of a connected data platform has been derived to satisfy the unusual requirements of this category of fleet management applications. The remainder of this book will therefore investigate the design of the individual system components and their interconnections. In particular, Chapter 4 will focus on the design of the data storage infrastructure, proving that this activity constitutes one of the pillars of the success of any connected vehicle platform.

4 Data Storage Infrastructure

Data analysis is the process of inspecting, cleaning, transforming, and modelling data in order to derive useful information, draw conclusions, and support decision-making. It can be a game-changer for many of today's businesses, as it helps streamline their processes, reduce their expenses, and boost their bottom line. The use of the conditional tense is intentional here, as these possibilities are currently hampered by the inability of companies to effectively manage the data at their disposal. Today, after all, our prime concern should not be the unavailability of data; the recent proliferation of monitoring systems and IoT devices has in fact led to an explosion in the volume of data, which mostly comes in the form of time series. This is data that offers remarkable fidelity, as it not only captures the final state of the observed assets but also their evolution over time. In fleet management, for example, time series allow continuous monitoring, meticulous description, and in-depth analysis of fleet vehicle behaviour. However, a greater level of detail does not necessarily translate into more informed decisions, unless it is accompanied by the effective use of the available data. This phenomenon is particularly evident in context-aware, real-time fleet management applications, where vehicle data does not exist in a vacuum but is deeply interwoven with context data. Further details on vehicle behaviour, in fact, become irrelevant without the ability to adequately combine this data with other information. Unfortunately, this is an atypical need that requires an equally unusual, and therefore non-trivial, data storage solution.

The objective of this chapter is to demonstrate that adopting an appropriate data storage strategy is decisive for the success of our platform. While the section on conventional database technologies illustrates existing database solutions, the section on designing the database argues for the need to depart from such technologies to adequately support context-aware, real-time applications.

4.1 CONVENTIONAL DATABASE TECHNOLOGIES

Databases are collections of data organised in such a way that it can be rapidly searched and retrieved electronically. Depending on their size, such collections can be hosted on a file system, in a computer cluster, or in cloud storage. Regardless of their location, data extraction is always entrusted to a dedicated entity, which is conventionally referred to as a database management system (DBMS). Still, considerable confusion hovers around the real meaning of the term database, as it is frequently used in a broad sense to refer to any of the database, the DBMS, or the associated application. Even looser is the definition adopted outside the IT world, whereby any collection of data, like spreadsheets, is labelled as database. Spreadsheets are undeniably an excellent tool for numerical processing, but when faced with large volumes

DOI: 10.1201/9781003688549-4

of data, multiple users, and complex computations, they reveal all their inadequacies. Conversely, databases are articulate and versatile tools, capable of accommodating a wide range of file types and sizes; they provide sophisticated search and filtering capabilities – even concurrently to multiple users – and limit unauthorised access to stored records. Unquestionably, any company can create value with a database, since the ability to store, retrieve, and sort information inevitably makes it more efficient, helping the company compete in the marketplace. However, resorting to a database is neither a trivial nor a straightforward task. Business objectives, corporate assets, and the stakeholders involved are all factors that decisively influence the choice of the database. Consequently, each application tends to require its own data storage solution. The function of this section is therefore to present the two predominant forms of databases used in fleet management and the factors that drive their adoption. As we shall see, such databases handle information through diametrically opposed data models, which inevitably entail antithetical advantages and limitations.

4.1.1 Relational Database

Relational databases have dominated large-scale data processing applications since the early 1990s and, despite a slight decline in popularity, continue to do so: DBMSs such as Oracle, MySQL, and Microsoft SQL are often the obvious choice. The concept of relational database originates from the seminal work of Edgar F. Codd, who decisively deviated from the approach of navigational databases, where data was stored as a list of linked nodes and retrieved by reference hopping. According to his innovative vision, records were instead to be organised in a series of tables, each dedicated to a different entity, making it possible to search for information by content, rather than by following the links between stored objects. This approach soon revolutionised the entire data processing landscape, bringing countless benefits to a broad swathe of applications.

Relational databases are particularly suited to managing highly structured data, as they allow the relationships between different entities of a given application to be explicitly stored. They are therefore the perfect playground for data analysis activities, which can exploit these links to generate new knowledge. In fleet management, this feature proves invaluable for performance assessment activities, as it allows to arbitrarily aggregate, filter, and analyse clusters of data covering various entities, such as vehicles, their drivers, and the operations they serve. A relational database also features built-in mechanisms to eliminate redundancy and guarantee the integrity of manipulated records. This is vital for online transaction processing (OLTP) applications like vehicle operations management, as it would prevent, for example, the same vehicle from being simultaneously activated for multiple operations. Still, despite these advantages, many applications forego the trappings of a relational database and adopt what is widely known as a NoSQL model.

4.1.2 Non-Relational Database

Non-relational databases flooded the market at the beginning of the new century, although their existence dates back as far as the late 1960s. Their recent rise is certainly legitimate, dictated by the compelling need of new Internet giants such as Google and Amazon to handle previously unimaginable volumes of data. Their widespread adoption, however, has been more of a knee-jerk reaction, inspired by the prejudice against the limited performance of relational databases.

Non-relational databases are designed to store unstructured data, such as much of the information that flows tirelessly through the Internet. For this reason, they do not employ the tabular schema typical of relational databases but support a number of far more flexible storage models. For example, data can be stored in a series of documents devoid of any predetermined structure, which is ideal if the records feature a non-homogeneous set of attributes. In fleet management, the most common form of non-relational storage solution is the time-series database, which regards fleet vehicles as IoT devices and stores their data as timestamp/measurement pairs. Such databases are mainly used for near real-time monitoring of fleet status, as the presence of an implicit data structure guarantees a significant performance boost. Time-series databases are in fact able to timely ingest large amounts of data from a myriad of sources and to retrieve stored data just as efficiently, with extremely fast queries despite operating on a multitude of data points. They can also handle out-of-order and late-arriving data – a feature of utmost importance since vehicles may lack a reliable connection as they are regularly in motion. Non-relational databases, however, fail to provide all the advantages offered by their relational counterparts. Hence, the choice of a database should depend on the application to be developed; unfortunately, it often becomes a question of the volume of data to be managed.

4.1.3 How Data Volumes Influence Database Choices

The unrelenting growth in the volume of data is particularly difficult to perceive. To understand the magnitude of this phenomenon, it suffices to mention that, at the beginning of 2020, the number of bytes in the data domain was 40 times higher than the number of stars in the observable universe. It is an exponential surge in the amount of information produced every day, which inevitably places ever-increasing demands on the technological capabilities of our data storage solutions. It was precisely concerns about the ability to efficiently manage large bodies of information that led, over the past two decades, to the meteoric rise of unstructured and semi-structured databases, which pushed the relational model to the sidelines. In fact, one of the main arguments for adopting non-relational databases, especially for IoT applications, is their robustness when facing large data volumes. Admittedly, such databases lack many of the useful features of relational models, such as robust support for secondary indexes, complex predicates, a rich query language, and the ability to join tables. Unlike the latter, however, non-relational databases are scalable, which means that it is easy to increase their capacity without sacrificing performance. Consequently, as time series data accumulate rapidly, there is a common – and certainly legitimate – belief that relational databases are ill-suited for such workloads. As a result, the choice of the database to be employed often comes down to simplistic considerations of the volume of data to be stored. This is at the expense of the fact that this decision should instead be based on a multitude of factors, such as the nature of the data to be managed, the operations to be performed on it, the profile of its users, the requirements of the application, and even the available budget. Therefore, although this approach tends to be correct, it could lead to the adoption of a non-relational database solution even when such a technology is incompatible with the needs of the application to be built.

As hinted at above, the main reason for preferring non-relational databases when handling large volumes of data lies in their ability to scale efficiently. Two basic approaches to database scaling can be distinguished, similar in that they both involve adding computing resources to the data storage infrastructure, but profoundly different in terms of implementation and performance. While *vertical scaling* relies on increasing the power of the existing machines of the database (in the form of CPU or RAM), *horizontal scaling* consists of adding extra units to the database cluster, with each node handling only a portion of the overall dataset. The major drawback of vertical scaling is that it is limited by the capacity of individual machines. Hence, as the amount of data or requests to be handled increases, a plateau will inexorably be reached, necessitating the addition of new nodes. This approach is, however, notably difficult when it comes to a relational database, as, typically, there is no obvious way to partition its data. Such databases were in fact designed in an era when their data could be kept small, neat, and organised. As a result, splitting the records of a relational database across multiple machines implicitly raises additional concerns that were not originally foreseen, such as the need to ensure the integrity of tables in a distributed environment. In addition, the data partitioning has a detrimental impact on the performance of these databases: when records are spread over a number of nodes, joins between tables may require cross-server communication, which inevitably adds to the work of the CPU. Non-relational databases, instead, scale horizontally with minimal performance loss; they remove some of the intrinsic constraints of the relational model, such as consistency, allowing the same information to appear multiple times in the database. This allows records to be stored as entire units on a single machine and copied or distributed between nodes as needed, eliminating the need for join operations. Non-relational databases thus offer highly scalable performance – provided, however, that all the advantageous features of relational databases are sacrificed.

In essence, the current rationale of using a non-relational database for IoT projects, or in general for applications that foresee continuous growth in the volume of data, is largely justified. For most of these applications, in fact, the features provided by the relational model are mere frills. However, this principle does not apply to all applications, and context-aware, real-time fleet management stands as clear testimony to this. Indeed, if this approach were to be followed for the above applications, the choice would inevitably fall on a non-relational storage solution, given the need to store an ever-growing volume of vehicle data in the form of time series. Doubtless, such databases can adequately support real-time monitoring of fleet status, but, on the flip side, they would be incompatible with activities like performance evaluation and vehicle operations management. In this peculiar form of fleet management, in fact, vehicle data does not exist in a vacuum, but it is heavily interconnected with context data. It is therefore essential to maintain and exploit data relationships, which necessarily requires the use of a relational database. In addition, the ACID properties guaranteed by the relational model prove essential for handling the transactions that are behind operations like dispatch, which would be particularly challenging to manage with a NoSQL database. The use of a relational database, however, is likely to lead to a performance degradation that may prove unacceptable for the time-critical decision-making behind emergency vehicle operations. For further discussion of these issues, the reader is referred to the relevant source.

Context-aware real-time fleet management applications pose a rather unusual case, as the need to portray vehicle behaviour through time series data would require the use of a non-relational database. At the same time, the services built on such data would demand the distinctive features of the relational model. Traditionally, a compromise in either direction would be necessary, leading in both cases to intolerable limitations. This leads, by necessity, to the search for an alternative solution.

4.2 DESIGNING THE DATABASE

The database is the cornerstone of countless digital projects and IT services. This is not surprising, since the database is designed to support operations that prove central to most applications, such as data storage, manipulation, and visualisation. What is more striking is the poor design of numerous database solutions. This often results in data not being stored correctly, preventing the user from obtaining the desired information within a reasonable time, or even retrieving it at all. A solid database can thus be compared to the foundations of a building: without it, even the best constructed software would fail. Database design theory appears to be not fully understood by businesses, despite the importance of these components. In fact, as pointed out in the previous section, we often witness the race for the best database, as advocated by the plethora of tutorials scattered across the Internet. No such database evidently exists, as every application requires its own solution. Each database must, in fact, be designed according to the requirements it has to support; otherwise, it would not be able to serve the applications it is intended for. However, this principle seems to be occasionally forgotten, even when it comes to critical applications such as the management of emergency fleets.

The main objective of this section is to derive the fundamental design of a database that can adequately support the management of the English National Health Service (NHS) ambulance fleet. In particular, what follows aims to demonstrate that there are solutions, different from those traditionally adopted, that can fully and uncompromisingly meet the unusual non-functional requirements of such applications. Then, starting from the analysis of the system's functional requirements, this section will also identify the data to be stored, their attributes, and the relationships between them.

4.2.1 Relational Database for Time Series Data

Context-aware, real-time fleet management applications require (NFR1) a Connected Vehicle Platform (CVP) capable of collecting any information that is relevant to describe the state of fleet vehicles at any given time. As previously established, this non-functional requirement calls for two different types of data to be taken into account: vehicle data and context data. According to the findings in Section 4.1, these data types would traditionally be stored in two distinct databases, non-relational and relational, respectively. However, storing records in separate units would result in substantial downsides. For instance, if the valuable information on the various fleet-related entities were kept separate from vehicle data, the quality of decision-making and the speed at which decisions are made would be adversely affected. In fact, when records are distributed across different storage units, the user depends on third-party tools to combine this data and turn it into something meaningful. Consequently, obtaining timely, business-critical insights could become extremely difficult and laborious. Therefore, separating vehicle data from context data proves impractical, making the use of a single storage unit the preferred approach (NFR2). Notwithstanding this, as outlined in the previous section, neither a traditional relational database nor one of the various NoSQL models proves suitable for holding this heterogeneous set of data. In fact, both approaches entail a compromise that would prevent the database from adequately supporting the services it is designed for. Indeed, such a choice would conflict with another requirement, which states that (NFR4) the CVP shall employ a storage unit capable of efficiently processing situation data.

The optimal solution would be a data storage unit that combines the advantages of both technologies: in other words, a relational database that can scale efficiently. Albeit counterintuitive, several exemplars of such databases are readily available. This technology was pioneered by Google in 2014 with Spanner, a georeplicated relational database, followed by solutions such as CockroachDB and Yugabyte. These systems belong to the recent wave of NewSQL databases, storage units that offer high scalability while retaining the traditional ACID notion for transactions. They can thus offer performance comparable to that of NoSQL databases, without having to resort to application-level workarounds to preserve record consistency. This apparently unattainable blend of features can be achieved through sharding, a procedure that conveniently splits database records into disjointed subsets. Most database schemas can, in fact, be transposed into a tree-like structure where descendants have a relationship with the root. Tables can then be partitioned based on the attributes involved in these relationships, so that all data of a certain entity is co-located in the same partition, avoiding the need for distributed joins.

Among the NewSQL databases, TimescaleDB is by far the most suitable for handling the situation data of context-aware, real-time fleet management applications. It is a relational database purpose-built for time series, engineered on PostgreSQL, and available via free software or as a fully managed service on major commercial cloud platforms such as Amazon Web Services, Microsoft Azure, and Google Cloud. TimescaleDB implements a unique form of sharding, called adaptive time/space chunking. This is an approach to database partitioning that accounts for the characteristics of time series data and the operations that are typically performed on it, which are clearly different from those carried out on other types of datasets. Sharding is performed on two dimensions, time and device ID. This means that, for each device, records are sorted chronologically and divided accordingly.

The innovation of TimescaleDB lies in the hypertable, which behaves like a standard PostgreSQL table but actually plays a much more significant role, automatically splitting data into chunks, each covering a specific time interval. The peculiarity of this form of sharding is that only the most recent chunks will be retained in the database memory, as new measurements, received in chronological order, are likely to be 'appended' to these partitions. This prevents unnecessary swapping of chunks to and from the database disc every time a write query is due, which often results in a severe degradation of node performance or, in other words, reduced vertical scalability. This strategy also benefits read operations, as the records they typically require – the most recent ones – are already in memory. Horizontal scalability is instead entrusted to a distributed hypertable, which automatically splits data into chunks across multiple machines. As Figure 4.1 shows, the resulting architecture consists of an access node, which stores the database metadata and performs query planning across the cluster, and a set of data nodes, which store the actual subsets and perform local queries.

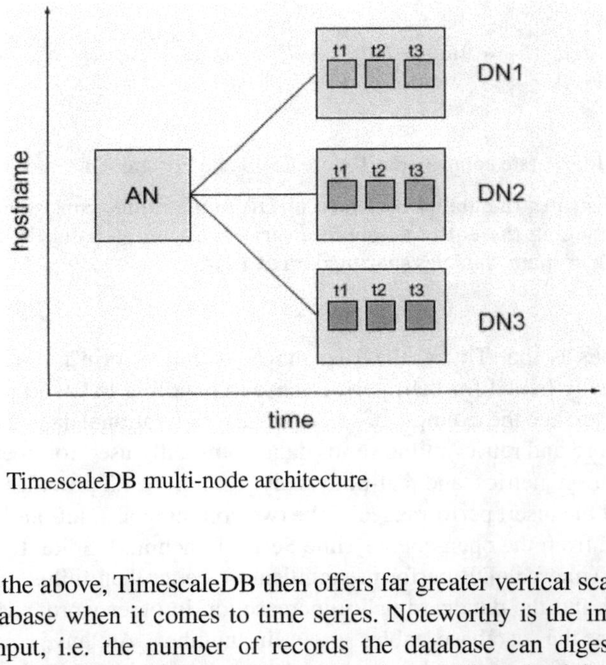

FIGURE 4.1 TimescaleDB multi-node architecture.

In light of the above, TimescaleDB then offers far greater vertical scalability than a relational database when it comes to time series. Noteworthy is the improvement in write throughput, i.e. the number of records the database can digest per second. TimescaleDB can load one billion database rows in one-fifteenth of the time required by PostgreSQL, as shown in Figure 4.2. This feature is crucial in the management of emergency vehicles, as it allows the collected data to be stored without any delay, regardless of the size of the fleet. Overall, all queries perform similarly or even better than vanilla PostgreSQL. Hence, if TimescaleDB were used for fleet management, the read queries used for live vehicle monitoring and the update queries employed in OLTP applications like dispatch would never be slower than those of a relational database. Moreover, TimescaleDB provides a number of useful time-oriented functionalities, such as time-bucketing, which collects records over arbitrary time windows, and last and first aggregates, which return the time extremes within a group.

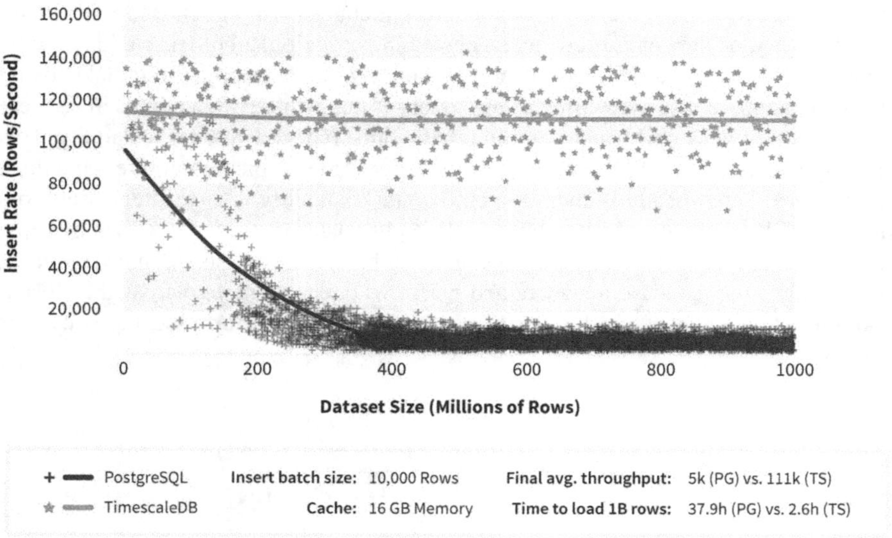

FIGURE 4.2 Ingest-rate comparison: TimescaleDB and PostgreSQL.

This figure illustrates the ability of TimescaleDB to guarantee consistent performance even when managing the entire ambulance service, assuming batch inserts of around 5000 rows, with as many vehicles sharing data at 1 Hz.

Less obvious is that TimescaleDB manages to outperform non-relational databases, perennially hailed for their performance in handling IoT workloads. To illustrate this, we can take the example of InfluxDB, a non-relational database specifically designed to store and retrieve time series data, commonly used for operations monitoring, application metrics, and real-time analytics. Figure 4.3 provides a quantitative comparison of the insert performance of the two commercial solutions against a standard workload from the open-source Time Series Benchmark Suite. It can be clearly observed that TimescaleDB performs significantly better than InfluxDB as the number of devices and the data cardinality increase, or, in other words, as the relationships between the database tables grow in number. As any emergency fleet management application involves a large number of vehicles and, above all, an unusually high level of interconnection between vehicle and context data, the choice cannot but fall on TimescaleDB. Similar results are obtained for read queries, with InfluxDB proving to be slower than TimescaleDB by up to tens of seconds for complex queries, further underlining that the decision to opt for a relational database is not so nonsensical.

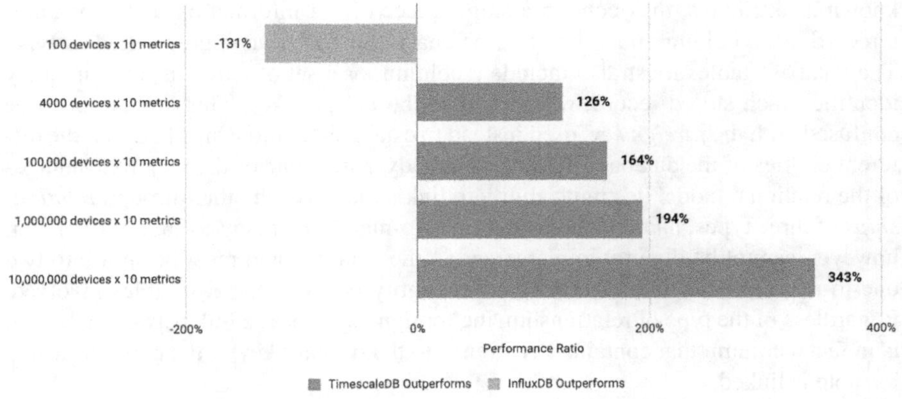

FIGURE 4.3 Performance comparison: TimescaleDB and InfluxDB.

For real-time fleet management applications, adopting the relational model brings further advantages, including the ability to use the Structured Query Language (SQL). TimescaleDB, in fact, leverages the power of standard SQL queries on time series, even at scale, permitting to query context data and vehicle data jointly and efficiently, as if they were conventional relational data. The opportunity to use SQL should not be taken lightly since, as the Stack Overflow Developer Survey points out, it is the third most common programming language today, used by more than half of the professional developers interviewed. In contrast, using a non-relational solution typically requires a different query language that, at best, resembles SQL. As a consequence, even business intelligence tools based on SQL, such as Tableau, become virtually inaccessible. TimescaleDB, on the other hand, guarantees native compatibility with such tools. Among these, the most relevant in the context of fleet management is definitely PostGIS, a Geographic Information System (GIS) database extender that allows spatial objects to be stored and manipulated like any other database object. It then becomes possible not only to store, interrogate, and analyse the position of individual vehicles by means of dedicated database types, but also to describe and evaluate the spatial relationships between the various points of interest, such as ambulance position and emergency location. PostGIS provides a great variety of spatial functions, such as distance, area, and intersection, which prove crucial for any emergency fleet management activity.

Having established that, for a context-aware, real-time fleet management application, the adoption of a relational database for time series provides numerous advantages, it is then necessary to determine the database structure. Details on the entities that form our TimescaleDB instance, their attributes, and the relationships among them will be covered in Subsection 4.2.2.

4.2.2 Database Structure

In a relational database, data is organised in tables, each of which is composed of a certain number of rows, representing records, and columns. Table columns, also known as attributes, thus contain a single piece of the information that constitutes a record: each column must therefore be characterised by an appropriate data type. The database tables must also include a column, or a set of columns, that uniquely identifies each stored record, referred to as the *primary key*. This key is not to be confused with the *foreign key*, used instead to establish relationships between the different entities of the database. In fact, as already stated, one of the major advantages of the relational model lies in its ability to link tables to each other through *relationships* of three types, namely one-to-one, one-to-many, and many-to-many. The latter, however, cannot be directly implemented in the database and must be split into two one-to-many relationships, creating a new entity between the two tables involved. Regardless of the type of relationship, the foreign key acts as a link between tables: it is in fact a column that contains a reference to the primary key of the entity to which its table is linked.

The above are the guidelines to abide by when designing a relational database. As such, the terminology set out above will recur repeatedly in the remainder of this subsection, which describes the structure of the individual tables and the relationships among them, as well as the considerations that lead to such design choices. The database schema is also included, summarising all information concerning the structure of the database. For further details on data types and for information on the various constraints the elements of the database are subject to, the reader is referred to Appendix A, which includes the source code for generating the database instance.

4.2.2.1 Measurement

A measurement represents the status of a fleet vehicle at a given moment in time. In other words, measurements constitute what was previously referred to as vehicle data: they are presented as a time series and provided with a *timestamp* that allows them to be indexed in chronological order. Since the device that generates the measurement must be taken into account, each measurement record is labelled with the *identifier of the dongle* it refers to. The latter, together with the timestamp, constitutes the primary key of the table. Indeed, the primary key could neither consist of the timestamp alone, as multiple dongles could produce different measurements at the same time, nor of the dongle identifier alone, as each device produces several measurements over time; in both cases, the unique constraint of the primary key would in fact be violated. Hence, a many-to-one relationship exists between measurements and dongles; this means that a dongle can produce several measurements, but a certain measurement can only belong to a single device. This relationship is expressed through the dongle identifier, which serves as a foreign key, being the primary key of the dongle table. Each record also features the *identifier of the vehicle* from which the measurement originates; for this attribute, the same considerations made for the dongle ID apply. The decision to not include either dongle data or vehicle data in the measurement table stems from the simple observation that this information, unlike time series data, is almost static. Including it would have meant replicating the same details for each measurement, resulting in an unnecessary increase in the space required by the database. These details, like the rest of the dataset, are in fact context data.

Each record must also include the geographical coordinates of the location where the measurement has been produced. For this reason, we have equipped the table with two numerical attributes – *latitude* and *longitude* – and a Spatial Reference Identifier (SRID), provided by PostGIS, to simplify all the geospatial operations the records might be involved in. In particular, we have opted for a *geometry* data type, which assumes that the ambulances are located on a Cartesian plane, rather than a geography data type, which assumes them to be placed on a spheroid instead. Notwithstanding slightly lower performance and negligible approximations, this guarantees the possibility of using geospatial functions that would otherwise be inaccessible; among these is the containment function, which is extremely useful in fleet monitoring to obtain the measurements of a particular area of interest.

Each measurement must also include the diagnostic data obtained from the OBD, specifically *vehicle speed* and *engine rpm* (revolutions per minute). The former is also used to estimate the *emissions* produced by the vehicle, information that is in turn included in the table as it is becoming increasingly relevant for performance evaluation activities. We could have included other raw OBD data or information that can be derived directly from it, such as the amount of fuel used by the vehicle; we could also have added data from accelerometers and gyroscopes, which are often used for driver behaviour monitoring applications. However, this information has been omitted, as its inclusion would have complicated the implementation of the CVP without bringing any substantial innovation.

4.2.2.2 Dongle

The dongle is the device responsible for collecting vehicle data, processing it locally, enriching it with GPS coordinates, and sharing it with the database. Being equipped with a network interface, our OBD-II devices feature a *MAC address*, which could be used as a primary key, considering that it is practically impossible for two devices to share an identical MAC address in the same network. However, our choice falls instead on a *sequence*, an integer number automatically generated by the database for each unique address, since it is shorter, simpler, and faster to handle by the DBMS. This table also includes an attribute for tracking the *status* of each dongle, which is useful, for instance, to record any device malfunctions. Finally, each record is characterised by a *buffer size* attribute, which is employed to determine the amount of data each dongle sends to the database per message; this value determines the number of measurements to be sent in a batch: if the buffer size is set to its default value '1', the dongle sends one measurement at a time. Considerations on the actual value of this attribute obviously depend on the specific requirements of the application, such as performance and cost constraints.

4.2.2.3 Vehicle

The vehicle is the object of fleet management activities. Hence, this table records all the data relating to ambulances, as our CVP is intended for the management of the NHS fleet; yet, this table only includes data that can be obtained, directly or indirectly, from the vehicle; this stems from our choice to implement a fully automated system, whose set-up does not require any human intervention.

The only parameter that is directly accessible from the on-board network is the Vehicle Identification Number (VIN), an alphanumeric fingerprint of 17 digits. Being the *VIN* unique for each vehicle, it could definitely be used as a primary key. As in the case of dongles, however, our choice falls on a numerical *sequence*, which is automatically generated by the database. The VIN can in turn be used to obtain other information: in fact, it is a standard string designed to encode details of the vehicle it refers to, such as its *make* and *model*, *year of manufacture*, *engine*, and *fuel type*. By cross-referencing the VIN with external databases, an even more detailed picture can emerge; for instance, we could acquire information about the year the vehicle was registered, its *gross weight*, and the size of its wheels. This information can be obtained using a VIN decoding service, which provides it at a very reasonable cost. A foreign key is also included, which is the *identifier of the* particular *emission model* that has been assigned to the ambulance. Each vehicle, in fact, depending on its characteristics, requires a different model for estimating the pollutants it produces.

4.2.2.4 Emission Model

Emission models are methods for calculating the level of pollutants produced by the various categories of vehicles, based on their speed and fuel consumption, emission factors, and the amount of traffic on the given type of road. Although a wide range of emission models are presently available, our choice has fallen on the average speed model provided by the UK National Atmospheric Emission Inventory (NAEI). Admittedly, this model does not represent the most appropriate option, as it generates estimates based solely on average speed, without taking into account any changes in the vehicle's operational mode. This makes it suitable for estimations on a road network scale, but inappropriate to be used on a micro-scale level. However, more accurate models, such as the instantaneous emission model developed by TU Graz (PHEM), require large budgets to be purchased and can prove particularly complex to implement. Since the sole intention is to demonstrate that extremely valuable information can be derived from plain OBD data, we decided to opt for an emission model that is clearly less accurate, but undoubtedly also less problematic.

The NAEI model categorises vehicles according to a number of key criteria, such as vehicle type, *weight*, *engine size*, *fuel type*, and *emission class*. As a result, the emissions table uses one column for each of these attributes, with the exception of vehicle type, since the only category that is relevant to our application is that of light commercial vehicles, or LGVs. Therefore, each entry in the table is labelled with a unique combination of these attributes, which identifies the particular vehicle configuration the model instance refers to. Despite this, the primary key of the table is a numeric *sequence* automatically generated by the DBMS.

The emission model allows for the estimation of four different pollutants, namely *CO*, *HC*, *NOx*, and *PM*, through the use of a single function, which is parameterised differently depending on the pollutant. More details on the methodology used to estimate emissions are available in Appendix C. As a result, all records feature four additional attributes, each defined as an array of eight values, containing the factors to be used to parameterise the model function according to the class of the vehicle. The association between an ambulance and its model is obviously performed during the set-up of the dongle, depending on the characteristics of the vehicle, whose technical specifications are obtained through the VIN (Table 4.1).

TABLE 4.1
Emission Model Attributes for Vehicle Classification

Attribute\Value	0	1	2	3	4	5	6
Weight	Not valid	<2.5 t	2.5–3.5 t	N/D	N/D	N/D	N/D
Engine Displacement	Not valid	<1400 cc	1400–2000 cc	N/D	N/D	N/D	N/D
Fuel Type	Not valid	Diesel	Petrol	LPG	N/D	N/D	N/D
Emission Class	0	1	2	3	4	5	6

4.2.2.5 Staff

Staff refers to the personnel operating the vehicles of the NHS fleet. Even though it can be assumed that the staff members have a personal *number* provided by the NHS, an automatically generated *sequence* is still used as the primary key. Obviously, this table stores the personal details of the operators, such as their *first name*, *surname*, and mobile *phone number*. A *password* is also foreseen, which is used by the operators to access the various functionalities of the system. This attribute, for security reasons, is not kept as plain text but encrypted using a hash and salt algorithm via the pgcrypto module and stored in a dedicated column. Each operator also holds a specific *role*: this information is stored in a separate table, called vehicle role, linked to the staff table through a foreign key. Evidently, a many-to-many relationship exists between the staff and fleet vehicles, as several operators can operate on the same vehicle at the same time, and the same operator can be allocated to different ambulances, albeit not simultaneously. To keep track of these combinations, a cross-reference table, named shift, must be created; both vehicles and staff will be linked to it via a one-to-many relationship.

4.2.2.6 Vehicle Role

Vehicle role is the duty that a staff member can assume when operating a vehicle of the NHS fleet. Four different roles are foreseen, each of which is characterised by a *name* and an *acronym*, as already outlined in Subsection 3.2.3. In addition, each role is provided with a *competence level*, defined as an integer with possible values between 1 and 4, which also serves as the table's primary key. This attribute is used during the dispatch procedure to determine the cumulative skill of the crews operating the various vehicles: this assessment ensures that only ambulances with the appropriate competence level are assigned to a particular call.

4.2.2.7 Shift

A shift is the period of time spent by a staff member on a given vehicle. As such, each record of this table contains both the *identifier* of the *staff* member and that of the *vehicle* to which the shift pertains. This implies that both attributes are foreign keys expressing a one-to-many relationship. Although it is common practice to use the pair of the two foreign keys as the primary key in a cross-reference table, for consistency reasons, our choice has still fallen on a *sequence*. To keep track of the time window over which a shift spans, a *start timestamp* and an *end timestamp* are provided. The correct use of the end timestamp also makes it possible to keep track of the status of the shift status without resorting to a separate attribute. An open shift is in fact characterised by a null end timestamp: when a value is stored in this column, it can be assumed that the shift has been closed.

4.2.2.8 Call

A call is the emergency that fleet vehicles must serve. The creation of a new call record follows the reception of a call by the control centre and is based on the details that the call handler obtains from the caller. At first, a call record features a few attributes only, such as its *identifier*, automatically generated by the DBMS, a *start timestamp*, the location of the emergency, and some details about the patient. As in the case of measurements, the position is modelled through *latitude*, *longitude*, and a *geometry SRID* attribute. This information is not provided by the caller but obtained from the *destination address* by using the Google Maps APIs. Patient information, instead, includes the *NHS number*, *full name*, and *phone number*. The NHS number is allowed to be initially null, as this information might not be readily available. Each call is also characterised by additional data, which is stored in a separate table, named *call type*, and referenced via a foreign key. Additionally, each record provides *further details* that may be useful to the staff serving the call; since this information does not fit into a set of predefined values, the corresponding attribute has to be modelled as plain text with no fixed length. As for shifts, the status of a call is handled through the *end timestamp*, initially set to null. The same applies to *service mode*, which can only be determined when the call is closed; since it should only accept two values, see and treat and see and convey, as per Section 4.2.2, this attribute can be modelled as an enum type.

Since in this prototype version of the CVP it is assumed that each call is served by exactly one vehicle, the relationship between vehicles and calls is of type one-to-many. Therefore, this table includes the *identifier of the ambulance* that will serve the emergency, which acts as a foreign key; this parameter is null when the call is created and is only set when a vehicle is chosen to serve it.

4.2.2.9 Call Type

Call type summarises the information of each potential emergency. Each record in this table is uniquely identified by a *sequence* automatically generated by the DBMS, labelled by a textual *description* of the emergency, and characterised by a specific *priority level*. As for the competence level of vehicle roles, the priority level of an emergency is an integer with possible values ranging from 1 to 4, in order to allow a direct comparison between the attributes of the two tables. Unlike the competence level, however, this attribute cannot be used as a primary key, as several call types may exhibit the same priority level (Figure 4.4).

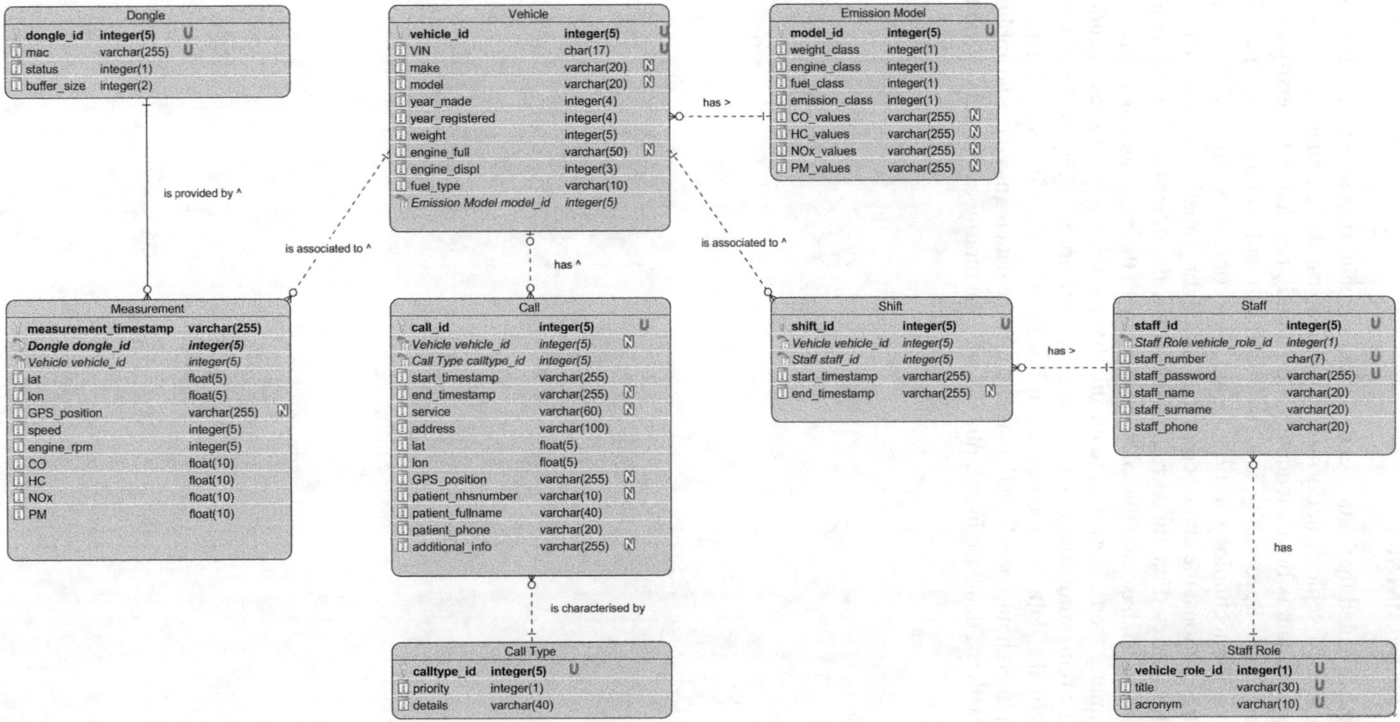

FIGURE 4.4 Database schema.

This diagram is not an exact representation of the database schema since the software used to produce it, Visual Paradigm, does not provide all the data types available in TimescaleDB.

4.3 CHAPTER SUMMARY

This chapter has revealed that database design is a crucial step in ensuring the success of the connected vehicle platform. Proper database design guarantees in fact that all the relevant data is both adequately stored and provided to the end-users in the correct time frame. Specifically, it is the non-functional requirements that drive the decisions on the technology to be adopted. For context-aware, real-time fleet management applications, the choice could not but fall on TimescaleDB, a relational database for time series. Functional requirements, on the other hand, have shaped the structure of the database and that of its entities, with their characteristics and the relationships among them. However, to ensure that the CVP functions as intended, the design of the remaining components cannot be overlooked. Chapter 5 will focus on the design of the data communication infrastructure: the aim is to determine which technologies should be adopted to assure that the data required by the application is shared with the database in compliance with the quality constraints currently in force.

5 Data Communication Infrastructure

Context-aware, real-time fleet management applications feature such unusual requirements that they require a data platform with a non-conventional design. As illustrated in Chapter 4, the need to maintain both vehicle data and context data in the same storage unit, as well as the need to ensure appropriate performance in manipulation of the resulting dataset, leads to the adoption of a database technology that is novel to the fleet management domain. Even the data communication infrastructure, which consists of the communication channels through which the different components of the CVP coordinate their actions and exchange data, involves design choices that are decidedly out of the ordinary.

As this chapter demonstrates, the peculiarity of the communication infrastructure of a data platform for context-aware, real-time fleet management applications lies in the unusual combination of communication technologies it relies upon. It employs the MQTT protocol for transmitting vehicle data at high frequency to the relational database – already an unconventional choice in itself – and uses HTTP for communicating context data, which requires, instead, a client-server interaction with the storage unit. This is a rather unusual configuration for a fleet management data platform, which typically relies on a single communication technology, chosen according to the services that the CVP is meant to provide. For instance, commercial platforms intended for traditional fleet management often use HTTP to share vehicle data with the database; others, such as those intended for fleet monitoring, use MQTT for a fast and reliable transmission of vehicle data to the platform back-end. Needless to say, our unconventional design is the offspring of the peculiar requirements of the applications the CVP has to handle.

The considerations outlined above, however, may prompt questions about the degree of innovation introduced by this work in the communication infrastructure. In fact, unlike the database, the data platform's communication layer is not built upon a novel or emerging technology, but rather on an unusual combination of solutions already established in the field of fleet management. Yet, it should be recalled – paraphrasing what has been stated at the beginning of this book – that the purpose of this work is not to describe an innovative product in all its individual components, but to demonstrate the importance of non-functional requirements in the design of the CVP: regardless of the level of innovation of the individual components, the platform will be innovative in its entirety, as it will prove capable of performance and functionalities that cannot be achieved with any other design. This is made possible by a particular, non-trivial, and definitely non-common combination of technologies, tailored to the requirements of the target applications. As such, it is not essential for all the technologies underpinning the data platform to be revolutionary.

DOI: 10.1201/9781003688549-5

The objective of this chapter is thus to demonstrate the importance of non-functional requirements in choosing the technologies that support the communication infrastructure of a CVP for ambulance fleet management. The section "Data Communication in Fleet Management" provides an overview of the communication technologies that could be used to implement the data communication infrastructure of a generic CVP; this passage is indispensable, given the absence, in the academic literature, of a comprehensive and reliable summary of both the advantages and limitations of such technologies in the context of fleet management. The section "Designing the Communication Infrastructure" draws on these findings to assign the correct technology to each of the communication channels of the platform, so as to meet the unusual requirements of context-aware, real-time fleet management applications.

5.1 DATA COMMUNICATION IN FLEET MANAGEMENT

Any form of communication between the nodes of a network is regulated by communication protocols – systems of shared rules that determine how messages must be handled. In the context of fleet management, the principal alternatives in terms of communication protocols essentially come down to HTTP and MQTT. There exist, of course, other technologies that could be employed to support communication between the various components of a CVP; among these are event-streaming platforms like Apache Kafka. However, event-streaming platforms cannot be subsumed under the category of communication protocols, as they not only regulate communication but also provide processing and analytical capabilities. Yet, although the spectrum of the technological alternatives that could be used is far from articulated, the situation that emerges from the literature turns out to be much more intricate.

As we've seen, the error that academic research seems to consistently fall into is that of neglecting the non-functional requirements of fleet management applications. This leads to studies that proclaim the need to use the same technology for every application, disregarding the fact that each CVP would instead require a distinct communication infrastructure. One of the most emblematic cases is the study by Aliwarga et al., which advocates the recourse to MQTT, as it guarantees a substantial reduction in communication latency compared to HTTP. This is a statement that is fundamentally correct and, for this very reason, fairly deceiving. Any reduction in communication latency is, in fact, extremely beneficial for services such as real-time vehicle monitoring. However, stating that this design solution is the ultimate choice, irrespective of the application to be implemented, is simply incorrect. Indeed, low latency does not bring any substantial advantage for services such as fleet performance evaluation analyses, where the processing of data is not required to be carried out immediately after data collection. Therefore, in this scenario, there is no evident reason to dispense with HTTP. This explains why numerous commercial players, like Geotab, continue to rely on HTTP to support their products, whose offerings consist, for the most part, of traditional fleet management services: innovating would not provide any substantial or immediate economic benefit. Another issue with these studies is that they tend to compare solutions on the basis of a single, arbitrarily chosen characteristic, which only adds further complexity to an already fragmentary literature. There are even examples of studies that compare these technologies without taking into account the constraints that naturally arise in the fleet management domain. P. S. Bist, for example, proposes the adoption of Apache Kafka for communicating vehicle data to the CVP, as it offers better scalability performance than MQTT. This argument is, however, visibly flawed, as it does not account for the fact that an event-streaming platform requires stable communication, which generally cannot be guaranteed in a fleet management application. For this reason, an event-streaming platform should not be considered as a mere alternative to HTTP or MQTT, but, as we shall see, should at most work alongside these protocols to underpin more complex fleet management services.

Hence, the aim of this section is to provide a critical evaluation of communication technologies such as HTTP, MQTT, and Kafka in the context of fleet management, analysing their advantages and limitations, and motivating the rationale for their use. Besides Kafka, the technologies investigated here are application-level communication protocols: the lower-level protocols on which they rely, like TCP, are not discussed, except for those aspects that can influence their behaviour. Moreover, since the emphasis of this section is on intra-system communication, the application-layer protocols that can be used to interface the CVP with external entities, such as vehicles, are excluded and will be reviewed in Chapter 6.

5.1.1 HTTP Protocol

The Hypertext Transfer Protocol, better known as HTTP, is an application-layer protocol intended for distributed, collaborative, hypermedia information systems. As its name suggests, HTTP is primarily used to transfer hypertexts, which are digital representations of traditional text structures enriched with references to other texts – called hyperlinks – that users can instantly access. In addition, HTTP can also be used for activities that go beyond its primary purpose, such as name servers and distributed object management systems. HTTP is therefore the staple of web navigation and, consequently, one of the mainstays of communication over the Internet: about 80% of today's websites rely on it.

HTTP is built on a request-response architecture based on the client-server model. It follows that any interaction between two network elements that involves HTTP is characterised by an explicit separation of roles between the entity that makes the request, known as the client, and the entity that serves it, referred to as the server. As such, not only the web browser that requests a web page is to be considered an HTTP client, but rather any device that requests the completion of a task on the network, such as the manipulation of a record within a database server. Hence, HTTP is not confined to web browsing alone, but lends itself to any web-based service, as long as it does not conflict with the applications' requirements. Among the various versions of HTTP, the one worth mentioning is HTTPS, which resorts to TLS (Transport Layer Security) to encrypt both requests and responses, proving far more secure than the original protocol. This HTTP variation ensures website authentication, privacy protection, and the integrity of the exchanged data. It is now even more prevalent than HTTP, given the pressing need to be protected from man-in-the-middle attacks, eavesdropping, and tampering.

HTTP was the preferred choice for supporting data communication in traditional fleet management applications, at least in the early days of commercial platforms. In fact, over the first two decades of the new millennium, this protocol was the foundation of all the services that formed the core of the value proposition of giants like Geotab. We are referring to activities like fuel management and driver behaviour monitoring, which perform analyses on vehicle data alone to gain insights aimed at reducing fleet costs in the short-term. The rationale behind the adoption of HTTP is to be sought in the exceptional familiarity that developers had with the protocol, many of whom had implemented web solutions of one kind or another. Nowadays, however, HTTP is hardly the ideal solution, as its specifics make the protocol unsuitable to meet the requirements of next-generation fleet management applications, like high transmission frequencies and high reliability.

To explain the reasons for this inadequacy, we must first consider that HTTP rests on TCP. It means that, prior to any exchange of meaningful information, a connection has to be established between the two entities involved in the communication. This is achieved through a three-way handshake, namely by the exchange of synchronisation and acknowledgement messages between the client and the server. In the original version of HTTP, connections were short-lived and, as such, exceptionally burdensome, since a new TCP connection had to be created every time a request was to be sent, and then closed once the corresponding reply was received. Another factor to take into consideration is the considerable size of the HTTP message, which employs a particularly large header. The overlapping of these two conditions made the transfer of vehicle data particularly onerous, both in terms of response time and network bandwidth. To reduce the latter and, coincidentally, cut transmission costs, fleet management companies have been resorting to batch data transfers: measurements were collected at arbitrarily high frequencies, then grouped, and finally sent in blocks at intervals of a few tens of seconds, in order to reduce both the number of handshakes and the burden of message headers on the communication. This workaround works flawlessly for fleet performance assessment activities since the data analysis is not required to be performed immediately after data collection, but can be carried out at a later stage. In this scenario, there is no evident reason to use a higher transmission frequency. However, the same cannot be said for live monitoring and operations management activities, which require the transmission of data immediately after its collection, and thus demand radically different communication performances. A better performance is possible with HTTP 1.1, which introduces the notion of persistent connection, allowing a single connection to be used to send and receive multiple requests and responses. The pipelining model goes one step further, allowing multiple successive requests to be sent without waiting for a response, significantly reducing network latency. Figure 5.1 provides an overview of HTTP connection management.

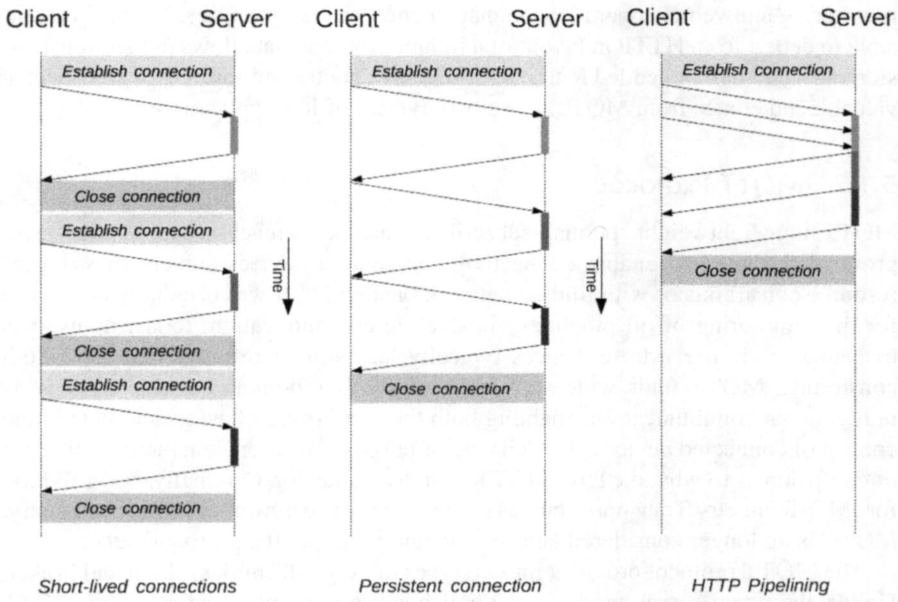

FIGURE 5.1 Connection management in HTTP.

The above approaches to HTTP connection management have the capacity to notably enhance the performance of fleet management applications. In fact, regardless of the volume of exchanges between the dongle and the server, such solutions can ensure shorter response times due to the reduced number of TCP handshakes. Yet, these approaches do not allow for any substantial reduction in the bandwidth used, which is largely conditioned by the HTTP message header. Therefore, the only viable solution to keep transmission costs low when using HTTP is to resort to low-frequency batch data transfers. However, this approach is not adequate for all fleet management applications. A further shortcoming of HTTP is its inability to natively guarantee message delivery: as a result, if dongles attempt to send messages without an active connection, the data shared via HTTP will never reach its destination. Regrettably, this eventuality is by no means uncommon: dongles, connected to an unreliable cellular network, often move through blind spots that can interrupt the communication between the vehicles transmitting the data and the CVP component responsible for storing it. When such an event occurs, the TCP connection must inevitably be re-established, regardless of the HTTP connection management method that is actually in place. In this circumstance, it is therefore crucial to resort to a buffer and a retransmission mechanism at the application-level, in order to ensure the delivery of those messages that would, otherwise, inevitably get lost.

HTTP, therefore, remains the most natural choice for traditional fleet management. However, this protocol suffers from a number of critical limitations that make it virtually unsuitable for supporting applications such as real-time vehicle monitoring and vehicle operations management. For these services, it would be sensible to depart from HTTP in favour of a lighter protocol that allows higher transmission rates and has embedded features for reliable data transmission. In this respect, a valuable aid comes from MQTT, a protocol typical of IoT applications.

5.1.2 MQTT Protocol

MQTT is a lightweight, publish-subscribe, machine-to-machine application-layer protocol designed to enable connectivity on remote devices subject to stringent resource constraints or with limited network bandwidth. It was originally conceived for the monitoring of oil pipelines via satellite communication; today, its use has extended to the IoT, whose devices typically face similar resource and bandwidth constraints. MQTT finds wide application in the IoT domain because it supports bidirectional communication, enabling both the monitoring of assets and the remote control of connected devices. This change of purpose has even generated an element of confusion as to what the term MQTT actually stands for. Originally, MQTT stood for 'MQ Telemetry Transport', but, as the protocol's use now goes beyond telemetry, MQTT is no longer considered an acronym and is simply the protocol name.

The MQTT protocol provides for two types of network entities, client and broker. Unlike the client-server model, the publish-subscribe pattern adopted by MQTT makes no distinction between senders and receivers, both identified as clients; any endpoint can assume both roles, even simultaneously. One of the peculiarities of the MQTT protocol lies in the absence of direct communication between the clients. All the messages shared on the network are collected by an intermediary, known as a broker, which forwards them to their intended recipients. This middleman is indispensable as messages do not contain any indication about their destination, but only information on the topic they refer to. To receive certain messages, a client must therefore subscribe to the particular topic to which these messages pertain. Several clients can subscribe to the same topic, and each client can subscribe to multiple topics at the same time (Figure 5.2).

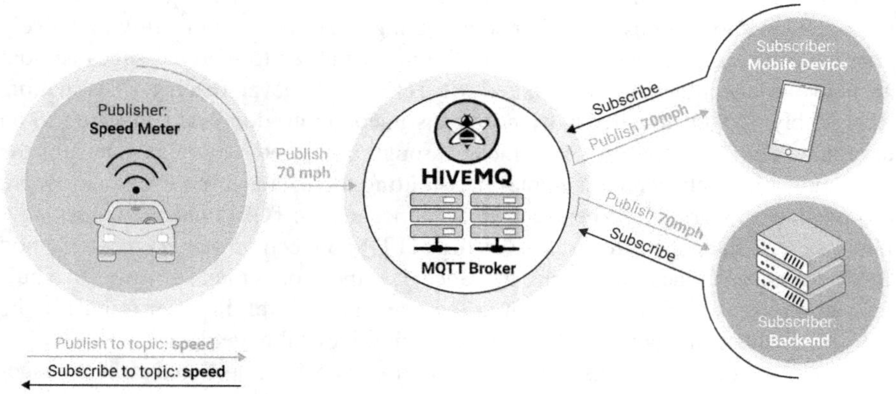

FIGURE 5.2 HiveMQ publish-subscribe communication.

The MQTT protocol typically runs over TCP/IP, or, alternatively, over other network protocols that provide ordered, lossless, and bidirectional connections. This means that, similarly to HTTP, a connection must be established via a three-way handshake before any meaningful information can be exchanged between an MQTT client and its broker. As for later HTTP versions, MQTT uses a single connection for transmitting multiple messages; in contrast to HTTP, however, the MQTT connection must be explicitly established by the client via a dedicated message, with the broker responding with an acknowledgement. This initial connection set-up increases the response time for sending individual MQTT messages, which may be even longer than the time required for sending an HTTP message. However, as the number of messages sent per connection increases, the average response time converges to values well below those obtainable with HTTP. This is due to the compact structure of the MQTT message, which, unlike HTTP, uses the binary format for its payload. There is also a notable difference in terms of headers, with those of MQTT being as small as two bytes, which ensures low bandwidth usage and low transmission costs. MQTT can then guarantee minimum overhead, low bandwidth usage, and, by reusing the same connection, low latency; this enables the high transmission frequency and the high responsiveness required by next-generation fleet management applications – unattainable with HTTP.

One of the central advantages of MQTT's broker-based architecture is that senders and receivers become decoupled, as there is no direct interaction between them. This *decoupling*, to be accurate, has several dimensions, each of which confers a different advantage in terms of performance. The first form of decoupling is *spatial*, as publisher and subscriber do not need to be aware of each other's IP address and port number. The inherent benefit is a preliminary level of security, as the communication endpoints are not addressable over the Internet. It is indeed the client, irrespective of its role in the communication, that establishes the TCP connection with the broker, meaning that no public Internet endpoint is ever exposed on the IoT device. In fleet management, it is thus practically impossible to infiltrate a dongle directly from the Web. Naturally, this is not sufficient to shield the vehicle from all forms of malicious attacks. In fact, since MQTT itself is not a secure protocol, it is advisable to adopt cryptographic protocols such as TLS/SSL, so as to endow the system with additional defences, including identity and authorisation checks, as well as payload encryption mechanisms. The second form of decoupling is *temporal*, by virtue of the fact that publishers and subscribers do not have to run simultaneously. The benefit is an increase in application resilience, as brokers become capable of holding messages for clients that are temporarily offline. Even clients can employ a similar mechanism, which is particularly useful when the broker becomes inadvertently unreachable due to an unstable connection. Both clients and brokers can in fact use a persistent connection, which allows messages to be stored locally until the recipient comes back online or the interrupted connection is restored. The level of reliability clearly depends on the chosen option of Quality for Service (QoS), which should be based on both the criticality of the application data and the stability of the connection. In critical applications such as emergency fleet management, this ensures that all the messages exchanged on the network via MQTT eventually reach their destination.

The communication that results from the use of the MQTT protocol can therefore be regarded as asynchronous, since both clients and brokers do not get blocked waiting for their messages to be processed by their respective communication partner. This means that, in fleet management, the use of MQTT makes it possible to develop extremely scalable CVPs, capable of both handling a large number of vehicles and supporting multicast communication, which would not be feasible with protocols like HTTP. Naturally, it is still necessary to ensure that the customer experience when interacting with the platform is not slowed down during peak usage periods. This can be achieved through the adoption of architectural solutions aimed at maintaining high service responsiveness: HiveMQ, which is the most popular commercial MQTT implementation in the automotive sector, employs a masterless cluster architecture that allows connections to be distributed over the various nodes of the cluster, depending on availability. Among the numerous strengths of the MQTT protocol is also the extensive support of its community. This is indeed one of the main factors behind the status of MQTT as the preferred protocol in the IoT domain, even at the expense of alternatives like CoAP and AMQP that, despite offering valuable operational capabilities, are far less supported. MQTT is, in fact, an open protocol with standard documentation, featuring a myriad of ready-made, open-source libraries for C/C++, Java, Python, and several other languages. All this, together with the availability of public MQTT brokers, minimises development costs and accelerates the implementation of any IoT project. A further advantage is the reduced power consumption of IoT devices. Yet, this benefit becomes marginal in the context of fleet management since dongles are not provided with a battery, but are powered by the car's battery via the OBD connector. MQTT is thus ideal for vehicle connectivity, not only in that it reduces bandwidth usage and communication latency but also because it provides a number of ancillary features that prove invaluable in a variety of fleet management applications.

5.1.3 Event-Streaming Platforms

An event-streaming platform is a highly scalable and durable system capable of aggregating large amounts of information from various sources in real-time. Such platforms allow the ingested records to be normalised, enriched, and filtered, and rules to be applied to the aggregated data to reveal relationships, trends, and patterns. The ultimate goal of an event-streaming platform is to capture business events as they occur and react to them promptly to deliver responsive customer experiences. As of today, the most popular platform is Apache Kafka, an open-source alternative to commercial solutions such as AWS Kinesis, Azure Event Hubs, IBM Event Streams, and Confluent Cloud.

Kafka is a distributed event-streaming platform built on a proprietary, TCP-based, publish-subscribe messaging protocol. Although it was originally conceived as a mere messaging protocol to support communication within LinkedIn, it has now evolved into a full-fledged streaming platform with built-in data processing capabilities, which finds application in a wide variety of other applications. Kafka currently manages passenger-driver matching for Uber, handles real-time analytics and predictive maintenance for the British Gas smart home service, and processes telemetry data in motion for several industrial IoT applications. Structurally, Kafka consists of a cluster of servers known as brokers, which are in turn composed of units called partitions. Here, messages, in the form of key-value pairs, are stored together with a timestamp for a limited period of time. This architecture provides Kafka with high-throughput, low-latency, and high-reliability. As shown below, the processes that publish messages are called producers, while those that process them are called consumers. As with MQTT, messages are forwarded to the appropriate processes by means of categories known as topics (Figure 5.3).

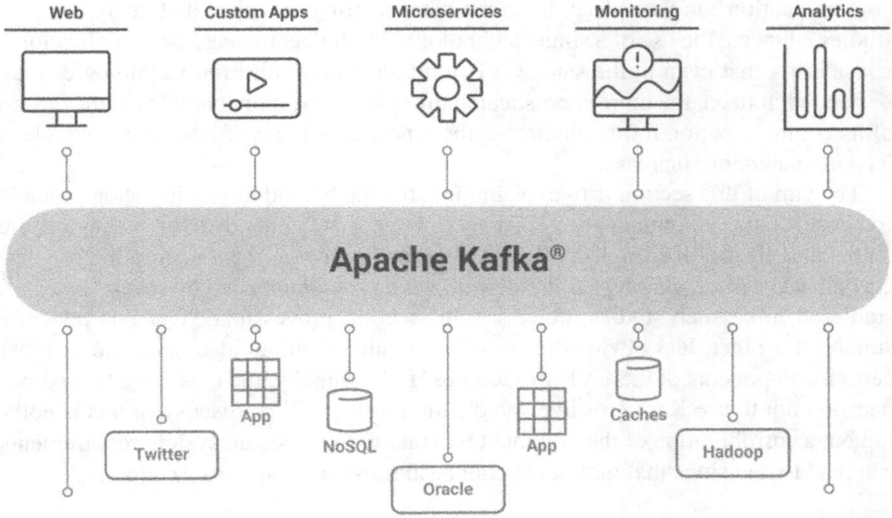

FIGURE 5.3 Apache Kafka: Processes.

As it turns out, Apache Kafka and MQTT share countless similarities. Even more significant are, however, their differences, which make them suitable for radically different purposes. MQTT, for example, is not designed for either data integration or data processing: when it comes to building high-performance data pipelines or integrating several business applications in real-time, the choice inevitably falls to Kafka. The latter, on the other hand, is not suitable for any IoT scenario, as it requires a solid network infrastructure: when it is necessary to enable communication in a non-wired or unstable environment, MQTT is then the obvious choice. Given its capabilities, Kafka can be used for real-time alerting – with the aim of generating automatic notifications upon the occurrence of certain events – for asset tracking, thanks to its filtering and map transformation functions, or as an intermediary between data sources for training models in machine learning applications. In fleet management, an event-streaming platform is particularly suited to the implementation of automatic notification systems for the detection of fleet-related events of interest, real-time fleet monitoring services, and predictive maintenance models, which allow proactive rather than reactive responses to the onset of faults.

5.2 DESIGNING THE COMMUNICATION INFRASTRUCTURE

Properly designing the communication infrastructure of a CVP for context-aware, real-time fleet management applications is by no means a trivial activity. In fact, for the overall system to function as envisioned, each of its components should be provided with a communication technology that allows it to interact effectively with the rest of the data platform. The technology to be deployed for each communication channel depends on the requirements of the target application: while functional requirements define the role of each component and, as a result, the type of interaction it requires, non-functional requirements may impose constraints on the way such communication can take place. It is then disconcerting to realise that many research studies advocate the use of a single technology for all fleet management applications, considering that even in the same CVP, multiple communication technologies may be needed. Indeed, the more sophisticated the system, the more complex is the design of its communication infrastructure, as the type and number of interactions between its parts inevitably increase.

The aim of this section is to explain the rationale behind the design choices made for each of the communication channels of our CVP, intended for managing the ambulance fleet of the English NHS. As will be discussed, the transmission of vehicle data takes place via MQTT, in line with what is recommended by several research studies. Unlike such studies, however, this section proves that MQTT is not only suitable for other, less obvious forms of communication but also demonstrates that certain components of the CVP still require HTTP. Finally, the reasoning behind our decision not to use Kafka for live vehicle monitoring will be discussed: this is nothing but a corroboration of the argument that taking into account system requirements can lead to decisions that may seem counterintuitive at first, cursory appraisal.

5.2.1 MQTT-Based Communication Interfaces

Context-aware, real-time fleet management applications require (NFR3) a CVP capable of capturing the necessary data for processing as soon as it becomes available. This requirement translates into interventions on both the structure of data producers and the configuration of the communication channels within the data platform, used to transmit all the gathered information to the remote data repository. To ensure that vehicle status information – hereafter referred to as measurements – is made readily available to the CVP, vehicle dongles must collect data at a sufficiently high frequency and transmit it equally frequently using a suitable network protocol. As shown earlier, MQTT is the most appropriate protocol for this purpose. Each dongle therefore runs an MQTT client module that sends the collected data to an MQTT broker, which stores the records in the database.

Along with a high transmission frequency of vehicle data, other requirements must be fulfilled to guarantee that the communication infrastructure of the CVP can appropriately support context-aware, real-time fleet management applications. The non-exhaustive list of requirements provided in Section 2.2.3 does not include, for example, the high degree of reliability required by vehicle data communication. In fact, in a critical application such as emergency vehicle management, it is not admissible that the messages shared on the network may never reach their destination, as every piece of information can be decisive in the decision-making process. Once again, MQTT proves to be the most appropriate solution, as it supports three different levels of communication reliability. As detailed below, the choice of the reliability level to be adopted depends not only on the criticality of the application data but also on other factors – such as costs and delays – that are perhaps less obvious but equally decisive.

The degree of reliability of the MQTT communication is embodied in a parameter known as Quality of Service (QoS), an integer value in the range of one to three. In emergency fleet management, the lowest QoS value should obviously be avoided, since it provides no guarantee that the exchanged messages will eventually reach their destination. The highest level of reliability is equally debatable since, while ensuring that messages eventually reach the intended recipients, it introduces both significant overhead on the network and substantial delays in the communication; this option relies, in fact, on a four-part handshake between the sender and the receiver to guarantee that the messages transmitted are received exactly once. The intermediate level of QoS, on the other hand, resorts to a single acknowledgement message to ensure that messages are delivered at least once to their destinations. This approach is arguably less onerous than the former, although it does introduce the possibility of messages being delivered multiple times. For emergency vehicle management, the most appropriate solution is then to opt for an *intermediate level of reliability* in order to be certain that the critical data transmitted by the dongles will reach its destination at least one time. The adoption of a higher degree of reliability is not advisable, as the overhead resulting from this approach would lead to significantly higher costs on the mobile network – given the high frequency of data transmission employed – and unnecessary communication delays. The chosen QoS value also requires the implementation of an application-level mechanism that assures that the messages already received are not processed again, but discarded.

Setting an appropriate level of QoS is, however, not sufficient to guarantee that the communication via the MQTT protocol is totally reliable. In fact, in the event that the connection between the sender and the receiver gets interrupted, all the messages that fail to reach their destination should necessarily be retained. This is possible by configuring MQTT clients so that they only establish *persistent connections* with the broker, in conjunction with a level of QoS greater than zero. Particular attention must be paid to the case of half-open TCP connections, in which one side of the communication continues to function normally as it is not notified of the failure of its counterpart. Here, the sender, unaware of the lack of a valid connection, continues to forward messages that will never reach their destination. Such a scenario is particularly common in fleet management due to the instability of the mobile network through which transmission between the dongles and the MQTT broker takes place. To alleviate this problem, together with persistent connections, MQTT provides a keep-alive mechanism that allows the nodes involved in the communication to determine whether their connection is still active or not. This measure is implemented by appropriately setting the keep-alive interval, which is a numeric value specifying the number of seconds that an MQTT client can be inactive and still be considered to be functioning correctly. If a client does not transmit any message during the keep-alive period, it will be required to send a ping packet to confirm its availability and simultaneously verify that the broker is still reachable; so, if a client does not exchange any message within one and a half times the keep-alive interval, the broker can then proceed with its disconnection. For emergency fleet management, we recommend a *keep-alive* interval of *10 seconds*. The adoption of a longer interval is discouraged because it would delay the retransmission of a large number of messages following the reconnection of the client, given that the frequency of data transmission is set to 1 hertz; shorter intervals may also prove counterproductive, as the physiological delays of the mobile network could generate false positives and lead to unnecessary client disconnections. Client reconnection is handled via an exponential back-off algorithm: the algorithm prevents broker overload by doubling the time between each unsuccessful reconnection attempt, from a minimum of 1 second to a maximum of 2 minutes, while adding a random delay to better distribute the synchronous reconnects of many clients.

Another non-functional requirement of paramount importance concerns the security of transmitted data. As detailed in Section 5.1.2, being MQTT not natively secure, the message exchange should be protected via *TLS*. This is a cryptographic protocol that resorts to a handshake procedure to establish a secure connection between the entities involved in the communication, guaranteeing that no attacker can intercept any part of the data transmission. Such protection is of great value in the context of emergency fleet management, as it ensures that critical data, such as the status and location of vehicles, cannot be intercepted by any malicious user. The use of MQTT over TLS has, however, a drawback, as security comes at a cost in terms of communication overhead. Yet, the burden of the TLS handshake becomes negligible as the number of messages exchanged over the same connection increases, as in the case of fleet management. Furthermore, techniques of session resumption can drastically improve communication performance, allowing a previously negotiated TLS session to be reused when reconnecting to the same server.

TLS also entails a significant CPU usage overhead, which is normally negligible on the broker but can be problematic on clients that are not designed for computationally intensive tasks. However, the dongles intended for this data platform, as we shall see in Chapter 6, are immune from such a problem: they are, in fact, endowed with greater computing power than the OBD-II readers that are currently on the market. This even encourages the implementation of additional defences, not included in the current design, such as authentication, authorisation, and payload encryption mechanisms, for whose details the reader is referred elsewhere. Indeed, this book does not claim to provide detailed information on all the requirements of a commercial CVP; therefore, details on strategies to meet secondary requirements like security and reliability are only provided if they can influence the attainment of the core requirements, without which the CVP would not function as intended.

Finally, a decision has to be made as to which commercial MQTT platform is to be used for measurement communication. Emergency fleet management applications require, in fact, that all the relevant information should be stored in a database for later retrieval. This is a rather unusual use-case for the MQTT protocol, which is typically employed to transmit data to the various clients for immediate consumption. The recourse to an MQTT platform capable of effectively handling data persistence is therefore indispensable. When using MQTT, the obvious solution to achieve data persistence is to set up a client that subscribes to the wildcard topic, so it can receive all the messages transiting the broker: this client can store data each time a new message is shared on the network. From a performance point of view, this strategy works extremely well with a low message throughput. However, when dealing with thousands or tens of thousands of publishing clients, as in the case of fleet management, it is likely that the client in charge of data persistence proves unable to handle such a hefty workload, ending up thwarting the throughput of the system. It should also be emphasised that the messages that the broker forwards to this client would necessarily transit the network, which would result in an unnecessary traffic increase. Surely, it is possible to run both the broker and the client on the same machine, which would overcome the traffic problem; however, the two entities would share system resources: the messaging overhead would then burden both modules, which is certainly not optimal. In view of the above, it has been opted for the use of *HiveMQ*, one of the few commercial MQTT implementations that is able to efficiently handle data persistence. HiveMQ, in fact, allows the broker's capabilities to be arbitrarily extended with additional functions via a plugin mechanism. It is therefore possible to write and hook custom code that enables the broker to store data in the database, eliminating the need for a dedicated MQTT client. This removes the previously mentioned performance bottleneck, since plugin executions are totally asynchronous.

Yet, the scope of MQTT cannot be confined to the sole transmission of measurements from the dongles to the database. In fact, its characteristics make this protocol ideal for informing the different data platform components about the outcome of the various decision-making activities that take place in the system. For instance, MQTT is used by the web app, downstream of the dispatching procedure, to communicate the details of an emergency to the vehicles that have been chosen to serve it. In light of the above, a dongle acts not only as a publisher but also as a subscriber, as it receives from the broker any information about the vehicle it is connected to. As can be seen in Section 6.2, the topic used for this communication reflects the unique identifier that is assigned to the vehicle in the database so that the dongle, when connected to the vehicle, can retrieve it. In this scenario, the advantage brought by MQTT is that the communication between the dispatching system and the ambulance can take place without the IP address and port number of the dongle having to be exposed on the network, thanks to *spatial decoupling*. Once received, destination data can be used by the dongle – in combination with Google Maps APIs data – to determine the best route to the patient, which can be displayed on the user interface of the dongle. Hence, route planning is performed without any input from staff, further reducing the response time to intervention.

Equally important is to notify the crew operating the ambulance of the dispatch that occurred. For this reason, even the phone application that is provided to staff members is equipped with an MQTT module. Every time an operator starts a shift on a certain vehicle via the application, the MQTT module subscribes to the topic associated with the vehicle, which is then withdrawn at the end of the shift. As a result, the web application becomes capable of autonomously notifying all the operators that are currently active on the ambulance that has been dispatched, regardless of their number, type, or location. The web app also shares over MQTT any information that is useful for the crew to properly handle the emergency, such as the destination of the call, the type and severity of the intervention, and details about the patient. Clearly, the exchange of messages between the web app and the various operators active on the dispatched ambulance takes place simultaneously, thanks to the message-level *multicast communication* enabled by MQTT, which would not be possible with client-server protocols like HTTP. The phone app also provides a function that allows operators to terminate the call their ambulance is serving once the intervention is completed. When this occurs, the phone app sends an MQTT message on the topic associated with the ambulance, so that both the dongle and the rest of the staff operating that vehicle can obtain this information. The latter is an asynchronous notification that allows the dongle and the phone apps to update their state without having to periodically check the status of the call in the database via a cumbersome polling activity. For a detailed description of vehicle dispatch, shift start, and call end operations, the reader is referred to the diagrams in Chapter 7. As with the dongles, the MQTT interfaces of the phone and the web applications use *TLS* and rely on *persistent connections*. Unlike the former, however, they employ the *highest QoS value*, as their communication is sporadic and the handshake overhead introduced by this reliability level is therefore negligible. Furthermore, resorting to this QoS level eliminates the need for application-level controls aimed at discarding the MQTT messages that have already been received, which could lead the notification recipients into an inconsistent state.

In practice, MQTT proves to be the ideal choice for a CVP for context-aware, real-time fleet management applications. In fact, it lends itself to several uses within the system, including some less obvious ones, as summarised in Figure 5.4. Nevertheless, certain interactions may still require a different communication model, such as the client-server model proper to HTTP, as discussed in Subsection 5.2.2.

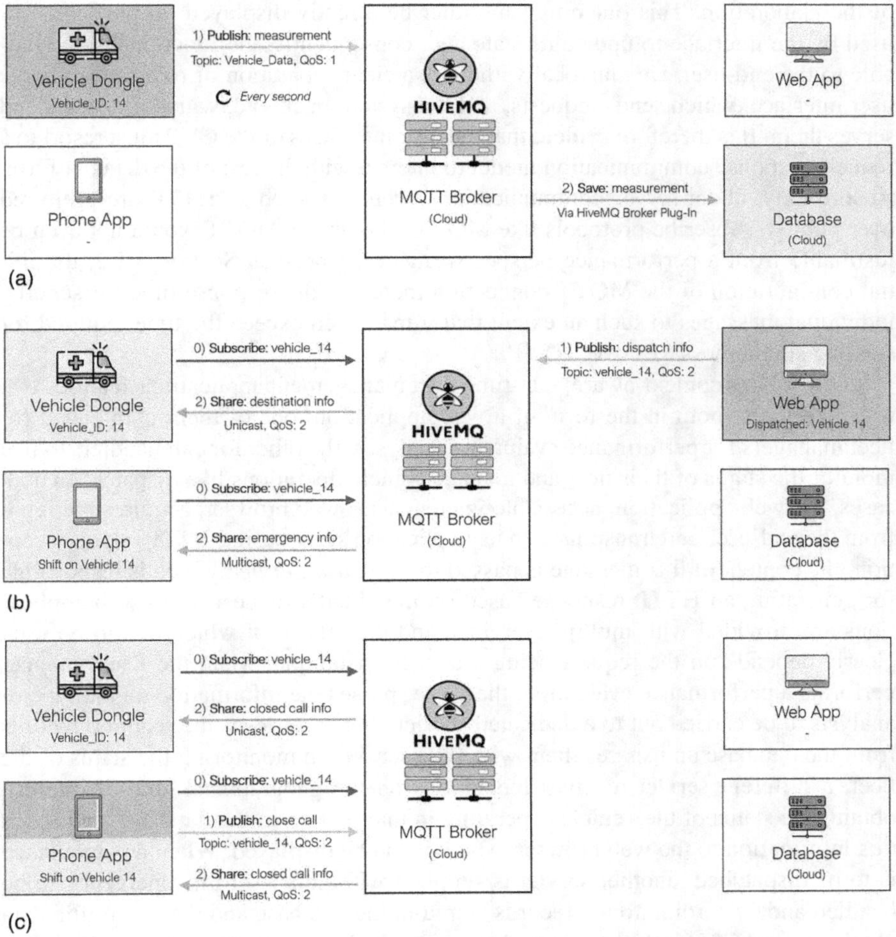

FIGURE 5.4 CVP component interactions via MQTT. (a) Measurement data transmission and storage. (b) Component notification upon vehicle dispatch. (c) Component notification upon call closing.

5.2.2 HTTP-Based Communication Interfaces

A CVP for fleet management applications is provided with a number of interfaces that allow its end-users to interact with the system and perform different activities, such as performance evaluation analyses, fleet monitoring, and vehicle operations management. Any user interface, regardless of its structure, technology, and functionalities, operates according to an extremely consistent pattern: it acquires the input from the end-user, shares this input with the system so that it can initiate the appropriate processing, and waits to receive a response from it, which contains the outcome of the elaboration. This outcome can either be directly displayed on the screen or used by the interface to update its state and, consequently, the functionalities available to the end-user. Unequivocally, there is a clear separation of roles between the user interface, which sends requests, and the system back-end, which processes and serves them. It is therefore evident that the user interfaces of the CVP must resort to a request-response communication model to interact with the rest of the data platform. Accordingly, client-server communications protocols such as HTTP are preferred over publish-subscribe protocols like MQTT. The use of MQTT would not even be justifiable from a performance perspective: as evidenced in Section 5.1.2, the initial configuration of the MQTT connection increases the response time for sending individual messages to such an extent that it may even exceed the time required for sending single messages over HTTP.

Our CVP for context-aware, real-time fleet management applications features two user interfaces, both in the form of a web application; one of them is intended for fleet managers for performance evaluation analyses, the other for call handlers to live monitor the status of their fleet and manage vehicle operations like dispatch. In both cases, the web application, accessible via a generic web browser, acquires the input from the end-user and transmits it to its application server using HTTP. Upon reception, the content of this message is passed to a *Jakarta Servlet*, which is responsible for generating an HTTP response based on the client's request. Both web applications are provided with multiple servlets, and the choice of which one to be used clearly depends on the request being made. For example, when the fleet manager performs a performance evaluation, the server passes the information on the type of analysis to be carried out to a dedicated servlet, which retrieves the required records from the database and shares them with the user. When monitoring the status of the fleet, a different servlet receives indications on the geographical area of interest, obtains the status of the vehicles operating in that area from the database, and sends this information to the web browser, where it can be displayed. When an ambulance is to be dispatched, another servlet is supplied with details of the emergency to be handled and, according to the records stored in the database and the live traffic data obtained from Google APIs, selects the most suitable vehicle to serve the call; it then stores this information in the database and transmits the outcome of the dispatch to the call handler. For further details on the structure of the web applications and the role of the Jakarta HTTP servlets in the data platform communication, the reader is referred to the corresponding chapter and appendix.

The CVP also provides for secondary users, namely individuals who access some of the system's functionalities without, however, actively participating in the management of the fleet. The ambulance operators are a perfect example of secondary users for the ambulance management application, as they interact with the system despite not personally using any fleet management services. Because of this, the CVP is also equipped with a user interface for the ambulance staff in the form of a phone application. As with the web applications, the communication between the mobile phone of the staff and the back-end of the system takes place over HTTP.

As explained in detail in Chapter 7, the Android application that the platform is equipped with provides ambulance operators with a variety of different functions; among these is the sign-up function, used by the staff to create their personal profile. This function is responsible for acquiring the user's data from the interface, packaging this information into an HTTP message, and transmitting this message to the data storage unit, where the operator details will form a new database entity. Even the data exchange underlying the log-in function is based on HTTP; here, the protocol is used to transmit the user's credentials to the server, which will verify the existence of the staff member in the database and then share its unique identifier. This ID is employed by the phone app to configure the user's personal area, which allows the ambulance operators to access additional HTTP-based services, such as shift and call management. Hence, the need to equip the phone application with an HTTP client module. This module has been developed around the *Android Volley* library, whose use makes the communication over HTTP extremely efficient, as it guarantees, among other features, multiple concurrent network connections and automatic scheduling of network requests.

The exchange of data between the HTTP module of the phone application and the back-end of the data platform is entrusted to a software interface, known as *Application Programming Interface* (API), which receives requests from the ambulance operators and fulfils them by retrieving the desired resources from the database. This API has been designed in the REST (Representational State Transfer) architectural style, for details of which the reader is referred to external sources. The recourse to REST grants countless benefits, including extreme communication scalability. This is due to the fact that a RESTful API is stateless, as it does not store any information on the interactions between a server and its clients. A RESTful API is also particularly responsive due to its ability to cache resources, which coincidentally ensures savings in both server resources and network bandwidth. Finally, it should be emphasised that REST does not place any constraints on the communication protocol used to implement the API. However, since a RESTful API must operate according to the client-server communication model, the obvious solution remains HTTP. That being the case, each request that the phone application submits to the API will contain two pieces of information: the HTTP method being used, depending on the type of server processing to be performed, and the uniform resource identifier – or URI – which specifies the target resource. Once the server elaboration is complete, the API shares the resulting resource with the phone application in JSON format: basically, the content of the HTTP message is presented in a human-readable text format in the form of attribute-value pairs. The API has been implemented using *Express*, a web framework for Node.js. It is a rather minimalist framework that has been supplemented over the years by innumerable compatible packages, which allow developers to tackle virtually any web development problem: there are libraries for working with user sessions, cookies, user logins, URL parameters, security headers, and much more. Naturally, the choice to use Node.js in the first place is not random. Despite its single-threaded nature, Node.js is, in fact, one of the most popular run-time environments to build server back-ends; this is because of its asynchronous and non-blocking behaviour, which allows it to handle tens of thousands of simultaneous requests, held in an event loop. The result is an extremely scalable API, capable of easily adapting to increasing workloads. This approach also improves the response time of applications and increases their performance, thanks to its remarkable efficiency in caching data and queuing requests, factors that should not be underestimated in next-generation fleet management services. Yet, no software is perfect, and Node.js is no exception. In fact, its non-blocking input/output model does not perform as well when it comes to CPU-bound tasks, leading to slow processing and overall delay in the event cycle. However, this is a workload that is largely irrelevant to context-aware, real-time fleet management applications, which makes *Node.js* one of the most suitable choices for building the API.

Even dongles, despite not serving as user interfaces, require an HTTP communication channel with the API. In fact, before it can start transmitting measurements via MQTT, each dongle must undergo a set-up procedure, which involves an interaction with the back-end of the system in a client-server fashion. In fact, during its start-up, the dongle must obtain from the database its unique identifier and that of the vehicle it is connected to, which are used to label the measurements acquired from the OBD. The parameters of the emission model associated with the ambulance, used by the dongle to estimate the vehicle pollutants, are also returned. To enable communication with the RESTful API, each dongle has been equipped with an *Axios* module, a promise-based HTTP client for Node.js.

Finally, a mention should be made of some of the strategies that can be adopted to improve the security of the HTTP communication. First of all, it should be emphasised that the API designed for this CVP is internal, and therefore hidden from any external user. Should it be necessary to share the information managed by the CVP with strategic partners of the NHS or other tools within the medical community, the adoption of an external API should then be considered. Notwithstanding this, the use of HTTPS remains recommended in order to establish a secure connection between the entities involved in the communication. Equally important is the use of an API key, to ensure that only authorised clients can access the system data, and that of a firewall, to control incoming and outgoing network traffic; API requests should also be sanitised, unused ports blocked, and access to remote file sharing protocols removed (Figure 5.5).

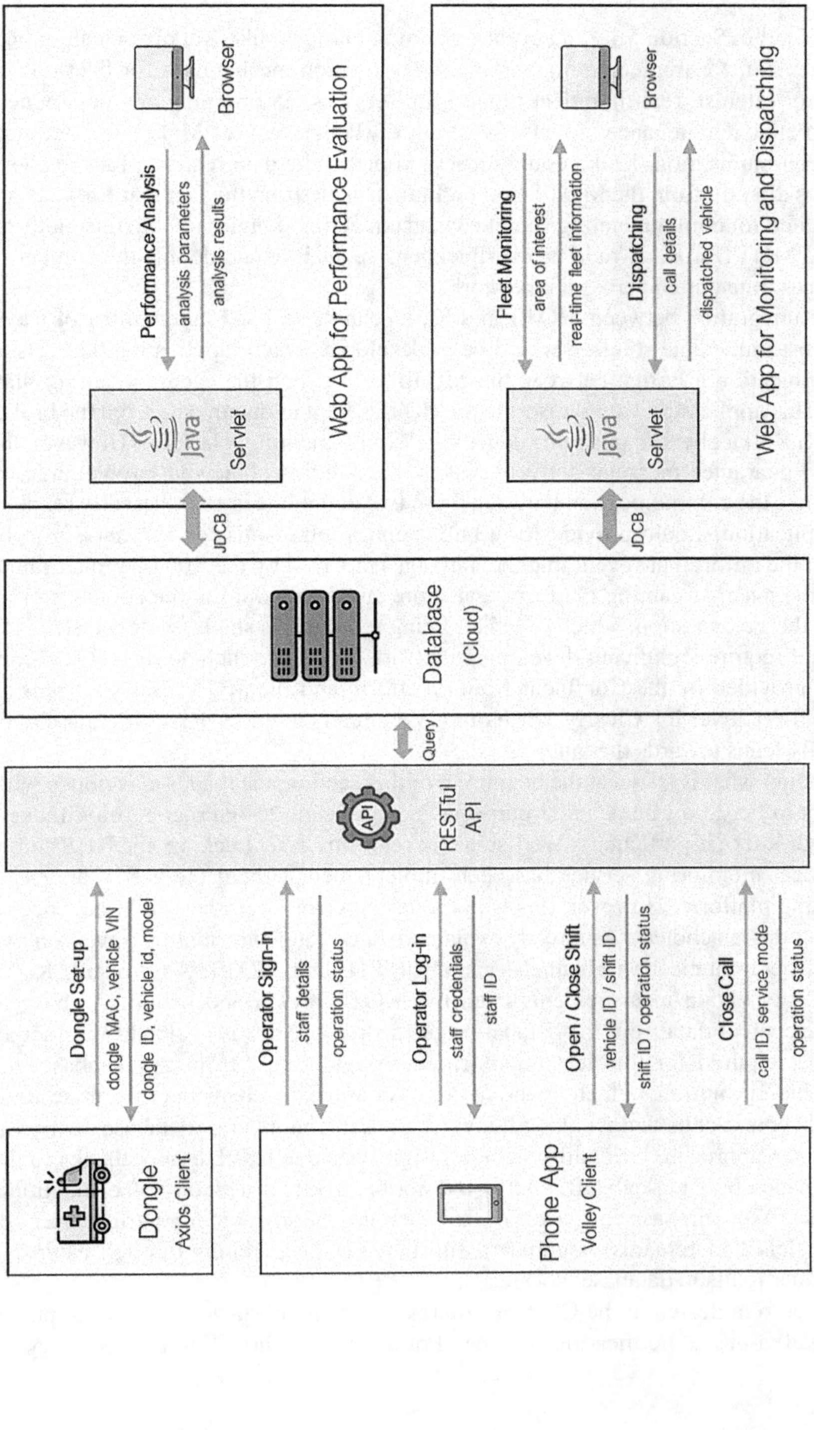

FIGURE 5.5 CVP component HTTP-based interfaces.

5.2.3 Event-Driven Notification Service

As detailed in Section 5.1.3, an event-streaming platform like Kafka is ideal, in fleet management, for implementing automatic notification mechanisms for fleet-related events of interest, real-time fleet monitoring services, and training and deployment of predictive maintenance models. Since our CVP already uses MQTT for measurement communication, Kafka could retrieve some of the data required for the above services directly from the MQTT broker. In fact, aside from the fact that Kafka is not well suited for communicating vehicle data, connecting dongles to both this platform and the MQTT broker would be overly expensive, as it would double the number of messages transmitted across the network.

The integration between MQTT and Kafka can be realised in a number of ways. The most immediate solution would be to develop a custom application that acts as communication a bridge between the MQTT broker and the event-streaming platform. This application would resort to an MQTT client to obtain data from the broker and to a Kafka client to push this data to the event-streaming platform. However, the need to guarantee message delivery makes this solution somewhat problematic. In fact, since the communication between these two endpoints is not inherently reliable, the application should provide for a buffer and a retransmission mechanism to be used in the unfortunate event that the messages received by the MQTT broker fail to reach the event-streaming platform. Far more straightforward is the adoption of an out-of-the-box solution, which – besides being reliable – is stable, widely tested, and does not require significant development effort. Examples include the MQTT connector provided by the Confluent Kafka platform and the *MQTT Broker extension plugin* from HiveMQ. Clearly, since our CVP already employs HiveMQ, our choice naturally leans towards the latter.

Reading what is stated at the beginning of this section, the reader may notice what appears to be, to all intents and purposes, a significant design incongruity. Indeed, although Kafka is particularly well-suited to real-time asset tracking, the NHS ambulance fleet monitoring service has been implemented without the use of an event-streaming platform. However, this is not a design error but rather a consequence of previous design choices. As already explained, fleet status information travels on two different communication channels, namely HTTP and MQTT. As a result, Kafka would only obtain measurements from the MQTT broker and would still have to retrieve context data from the database. For this reason, it was decided to extract all the data required for the fleet monitoring service directly from the database: the Kafka-based solution, in fact, would require a complex mechanism to integrate vehicle and context data, and would still involve the same number of database queries as the chosen approach. Evidently, such a design choice is feasible only thanks to the performance of TimescaleDB, and would not be practical with any other relational database. We emphasise, however, that despite the theoretical performance of TimescaleDB, it remains essential to substantiate these claims through extensive tests under realistic database workloads.

The current design of the CVP, by contrast, poses no obstacle to the development of a Kafka-based notification service. For example, should it be necessary to implement an automatic notification service for vehicle malfunctions, Kafka would only require an episodic interaction with the database. Indeed, all the information that is required to identify the onset of a vehicle fault is contained in the measurements shared via MQTT. Therefore, Kafka would only interact with the database after intercepting a fault, with the purpose of obtaining the records that are needed to contextualise the status of the affected vehicle. Such a service could be useful, for example, to restart the dispatch for the call that was being served by the now broken-down vehicle. Finally, the CVP could utilise Kafka to support the development and deployment of a predictive maintenance model: in fact, such a platform, on the one hand, would handle the model training using large volumes of data in motion and, on the other hand, would enable their deployment in real-time. However, while acknowledging both the feasibility and the advantages of implementing a predictive maintenance model in our system, we refrain from providing any indication on the integration between the CVP and Kafka for such applications. In fact, the interactions between Kafka and the various system components in support of a machine learning application are potentially very complex and would require in-depth investigations that are beyond the scope of this research project. In conclusion, a detailed representation of the interactions between Kafka, the MQTT broker, and the system database is presented below (Figure 5.6). The revised model of the CVP architecture, which incorporates all the findings presented in this chapter regarding the communication interfaces, is instead shown in Figure 5.7.

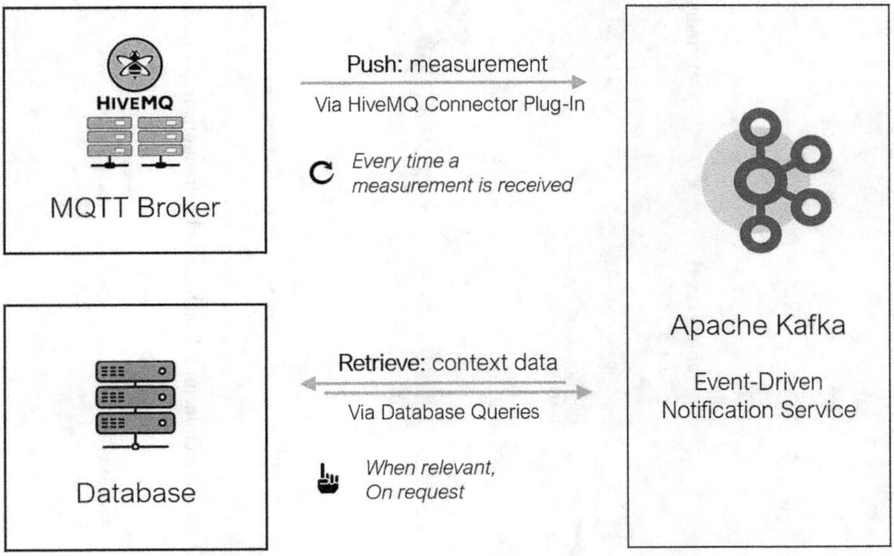

FIGURE 5.6 CVP: MQTT-Kafka integration

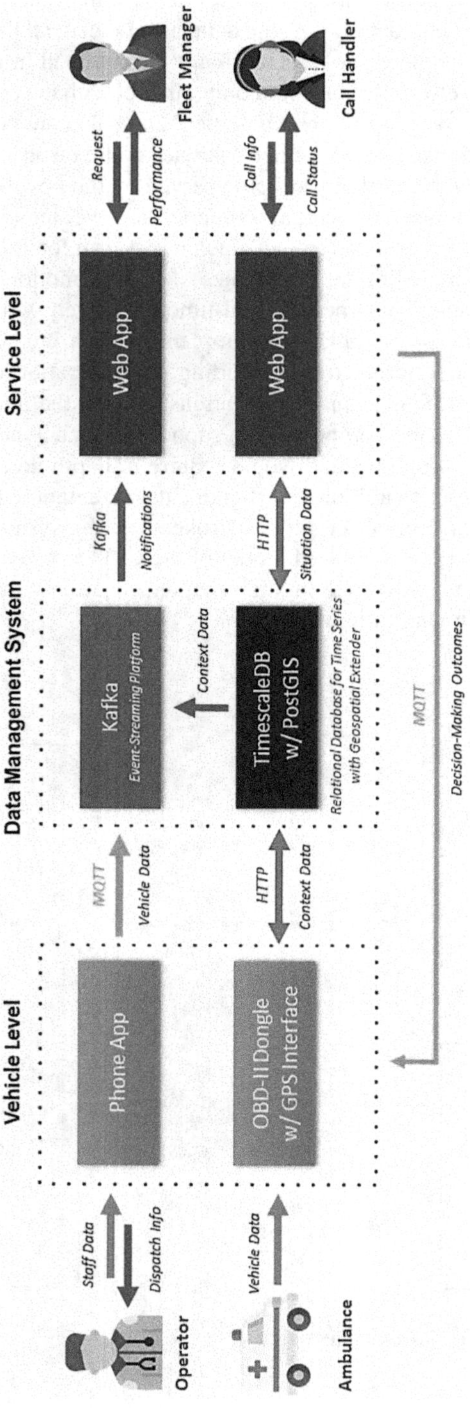

FIGURE 5.7 CVP architecture: Database and communication channels.

5.3 CHAPTER SUMMARY

This chapter has demonstrated the centrality of system requirements in the design of the communication infrastructure of a CVP for fleet management applications. In fact, any decision regarding the communication technologies to be adopted cannot stem from a cursory analysis of their characteristics, but rather from an assessment of their ability to meet the constraints that exist on the applications that the CVP has to manage. For this reason, in order for an ambulance fleet management application to function as intended, the data platform should make use of both the MQTT protocol, which guarantees frequent and reliable transmission of vehicle data, and HTTP for sending, manipulating, and extracting context data to and from the database. Moreover, precisely because of system requirements, an event-streaming platform should not be used for the real-time monitoring of the emergency fleet, but be relegated to other uses, in spite of the fact that the monitoring of assets is definitely within its capabilities. Having outlined the structure of the core elements of the data platform, it is now possible to proceed with the last phase of its design, which concerns its interfaces. Chapter 6 will cover the design and implementation of the dongle, which is the interface used by the CVP to collect data from vehicles. As with the other components of the data platform, we will demonstrate, once again, that the design of the dongle must necessarily take into account the requirements of the applications that the CVP is intended to support.

6 Vehicle Interface

The aim of this chapter is to describe the design of the vehicle interface, which is the data platform component responsible for obtaining data from the vehicle and making it available to the rest of the system. This task might seem rather simple, since the modern car is equipped with an astounding number of sensors, capable of generating a constant flow of data: according to a recent McKinsey estimate, this data amounts to 25 gigabytes per hour. It is indisputable that such a volume of data is only bound to grow, as the progressive shift to the Connected and Autonomous Vehicles (CAVs) paradigm will lead to an increasing reliance on digital electronics. It is therefore legitimate to look at vehicles as a source of big data on wheels. Unfortunately, despite its potential, vehicle data is not made readily accessible to unauthorised third parties for application development. In fact, in the automotive world, we are still faced with the age-old issue of vehicle data ownership, with the Original Equipment Manufacturers (OEMs) – proprietors of the data – unwilling to allow its monetisation. In this respect, the absence of a standard design for the on-board electronics plays to the advantage of the manufacturers: since data circulates in cars in a manner that is arbitrarily established by the OEMs, not only is an ad hoc solution required to extract data from each vehicle, but the participation of the manufacturers is also required. Today, only a limited subset of the data generated by vehicle sensors remains freely accessible: we are, of course, referring to the vehicle's diagnostic data set, which can be retrieved via dedicated protocols. Fortunately, this set of data is more than sufficient to support fleet management applications.

As can be inferred from the above, our CVP for context-aware, real-time fleet management application employs a vehicle interface in the form of an OBD dongle – a plug-and-play device that connects to the diagnostic port of emergency vehicles. This component is by no means ground-breaking, as it already underpins the vast majority of fleet management applications that are currently on the market. However, echoing what was stated at the beginning of Chapter 6, the aim of this work is not to build a data platform that is revolutionary in all its components, but to design a solution that is innovative as a whole. To achieve this, it is crucial to tailor each data platform component to the requirements of the applications that the CVP is meant to support, regardless of the level of innovation that might follow.

The objective of this chapter is to demonstrate the importance of non-functional requirements in the design of a vehicle interface for context-aware, real-time fleet management applications, ensuring the proper functioning of the data platform. "The Role of Diagnostic Data for Services" establishes the need for a dongle based on the ISO 15765 protocol, despite the fairly wide range of approaches available to obtain vehicle diagnostic data. Indeed, a device based on any other protocol within the OBD-II suite would not be as reliable for an emergency fleet; similarly, dispensing with the dongle in favour of an automotive data service provider would neither allow

 DOI: 10.1201/9781003688549-6

control of key parameters, such as the frequency of data acquisition, nor enable two-way communication with vehicles, which are core features for the target applications. Both alternatives, needless to say, would compromise the functioning of the overall CVP. Furthermore, the fact that the vehicle interface that results from this project is not technologically pioneering does not mean it lacks innovative features. "Designing the Vehicle Interface" demonstrates the need to depart from both commercial OBD dongles and the interfaces proposed in the academic literature, in terms of both hardware architecture and software structure, to achieve a design that meets the unusual requirements of context-aware, real-time fleet management applications.

6.1 THE ROLE OF DIAGNOSTIC DATA FOR SERVICES

The increasing use of embedded electronic components offers numerous advantages, including the ability to equip vehicles with self-diagnostic and reporting capabilities that enable them to readily notify what happens on board. Vehicle diagnostics, which originated as a small set of OEM-specific tests to detect basic car faults, now comes in the form of a sophisticated, standardised approach capable of detecting hundreds of failures. This evolution was driven partly by OBD regulation, partly by the OEMs' need to provide technicians with a common framework for performing complex inspections on vehicle engines and transmissions. Consequently, modern vehicles maintain open access to certain data, which can be retrieved simply by leveraging the diagnostic protocol supported by the vehicles – that is, the set of conventions governing diagnostic communication with a dedicated electronic control unit (ECU). The question remains, however, whether such data can legitimately be used for service development [15].

Diagnostic protocols have given anyone – even outside the circle of technicians – the ability to access information on vehicle status. This is achieved through the use of aftermarket diagnostic devices, known as OBD-II dongles. The vehicle telematics market is, unquestionably, extremely lucrative: suffice it to say, its size is projected to reach $779.98 Billion by 2030. Naturally, this has given rise to concerns about the undue monetisation of vehicle data. Indeed, even though a person owns a car, the rights to its data remain in the hands of the manufacturer; nevertheless, the use of a diagnostic protocol cannot be prohibited. This has prompted industry personalities to state – arguably to discourage the proliferation of OBD-II devices – that their use while vehicles are in motion poses a significant safety hazard. For example, Christoph Grote of the BMW Group made such a statement in 2017. However, this claim remains unsubstantiated, as evidenced by the countless fleets that use such devices safely. So, rather than a safety-related technological matter, this seems to be a legal dispute with significant economic implications. The messages of a diagnostic protocol have, in fact, a low priority on the communication network of the vehicle, and would therefore be discarded in the rare event of a network congestion, so as to prioritise information that is critical for the correct functioning of the car. However, since they do not exhibit a fully deterministic behaviour, diagnostic protocols should be avoided for the development of safety-related applications. Apart from this, though, a reliable dongle poses no more risks or hazards than any other vehicle interface, which inherently provides an access point to the on-board electronics. Further confirmation of this came in 2018 from the main preparatory body of the EU Council – COREPER – which recognised the need to maintain an open access to diagnostic data to support the development of services and applications in the aftermarket, which will be granted even while the vehicle is in motion.

As just explained, diagnostic data proves to be a legitimate dataset for the development of services in the aftermarket. However, the dependability of the OBD as a tool for supporting fleet management applications has yet to be demonstrated. In this regard, Section 6.1.1 will show that, on the one hand, this suite is supported by practically every vehicle in modern emergency fleets and, on the other hand, that its protocols can provide access to all the data required for their management. Section 6.1.2 will explain the reasons for choosing ISO 15765 over the other diagnostic protocols of the suite. In conclusion, Section 6.1.3 will prove that resorting to a vehicle data service provider such as Otonomo, while allowing access to diagnostic data without the use of a dongle, is decidedly impractical for the vast majority of fleet management applications.

6.1.1 Dependability of the OBD Suite for Fleet Management

Although frequently misused to refer to the diagnostic connector that most cars are equipped with, the on-board diagnostics (OBD) are a set of norms that establish the standard procedures to be used to analyse the status of electronic components in vehicles. It is a suite of protocols that covers every aspect of the vehicle diagnostics, from the structure of the diagnostic plug and its pinout to the electrical signalling protocols accessible through it, as well as the message format to be used to obtain diagnostic data. To understand the extent to which the OBD suite has penetrated the automotive market – and thus the degree of coverage it provides over emergency fleets worldwide – it is crucial to consider the regional variants of this suite. The latest version, OBD-II, to which reference is typically made, applies only to the North American market. Other countries enforce different standards, such as EOBD in Europe, JOBD in Japan, and ADR 79/01 for vehicles intended for the Australian market, to name a few. While equivalent to OBD-II in terms of capability and standardisation, these protocols came into effect at different points in time. The timeline for the adoption of the OBD in the various countries is shown in Figure 6.1; as hinted at earlier, these countries do not represent the locations where the vehicles are manufactured, but the markets they are intended for.

FIGURE 6.1 Timeline of adoption of the OBD suite.

In light of the above, resorting to the OBD suite not only constitutes a legitimate approach to accessing vehicle data but also a fairly dependable strategy, as these protocols are currently widespread in all the major automotive markets around the world. Yet, it could be argued that this suite is not completely reliable, since only the vehicles marketed in the last decade or so are supposed to be OBD-compliant: this is, however, of little relevance in the context of emergency fleet management, where vehicles are typically no older than 5 years, as outlined in Section 3.1.2. Presently, the sole criticality concerns the lack of coverage of the electric vehicles (EVs); in fact, while the OBD suite regulates hybrid electric vehicles (HEVs), to date, there is no legal obligation for an EV to comply with any diagnostic standard. However, a firm step in this direction should be expected in the near future. Regulators like CARB have already declared their intention to start monitoring the health of electric propulsion systems; added to this is the more veiled, but not less sincere, desire of OEMs to adopt a common diagnostic strategy to all vehicle types, which would guarantee a substantial reduction in costs.

What remains to be proven is that the OBD suite can provide access to all the data that is needed to properly handle context-aware, real-time fleet applications. Clearly, the amount of information accessible via OBD has grown enormously since its introduction in the 1980s. Early versions of on-board computers were capable of detecting the onset of a fault – signalled by a malfunction indicator light (MIL) – but could not provide details on the true nature of the problem. Today, it is possible not only to obtain detailed information on the failure of various vehicle components, but also to monitor their behaviour in real-time. The SAE J1979 standard specifies the exhaustive set of information that can be retrieved from passenger vehicles via the OBD. For heavy-duty vehicles (HDVs), the reference standard is SAE J1939, which employs a connector and diagnostic procedure different from those used in passenger vehicles. The latest version of SAE J1979 defines ten different services to access diagnostic data, three of which prove particularly well-suited to fleet management:

Mode 0. This service provides real-time information on the status of the electronic components that undergo mandatory emission controls. Data can be requested by forwarding a Parameter ID (PID), which prompts the diagnostic ECU to transmit the corresponding information. Much of the data required for fleet management can be retrieved via this service. This includes *engine load*, *engine speed*, *throttle position*, and *odometer* data, which can be used for fleet performance evaluations. OBD parameters, such as *engine coolant temperature*, *air flow*, and *fuel trim*, even allow the estimation of parameters that are not directly covered by the standard, such as *fuel consumption* and *emitted pollutants*. Fleet monitoring and vehicle operations management activities can leverage, instead, real-time data on *vehicle speed* and *fuel level*.

Mode 03. This service provides access to *diagnostic trouble codes* (DTCs), which the diagnostic ECU generates when it detects a malfunction or any other vehicle problem. This data can be used for both traditional fleet management and event-based notification services, as well as for the training of predictive diagnostic models.

Mode 09. This service grants easy access to the *VIN*, which encodes information such as the *make* of the vehicle, its *model*, *fuel*, and *engine type*. This information can be used for an initial, vehicle-level contextualisation of the car's measurements.

The OBD, as just seen, proves to be a highly effective and readily available tool for developing emergency fleet services. Indeed, it not only provides access to all vehicle data needed to successfully manage an emergency fleet, but is also legitimately available for virtually any emergency vehicle worldwide. All that remains is to choose the diagnostic protocol to use, as the OBD suite offers no fewer than five options.

6.1.2 Reasons for Using ISO 15765 to Access Vehicle Data

In accordance with the OBD-II protocol suite, access to a passenger vehicle's diagnostic data is to be performed via a standard hardware interface, formally known as the J1962 connector, but more commonly referred to as the OBD port. This 16-pin connector comes in two versions: *Type A*, fitted on 12-volt vehicles, whose pinout is illustrated in Figure 6.2, and *Type B*, mounted on 24-volt vehicles. The SAE J1962 standard regulates all aspects of the connector, including its location: it must be positioned no more than 0.61 metres from the steering wheel, unless the manufacturer is granted an exemption, in which case it may be placed outside this range but still within the driver's reach.

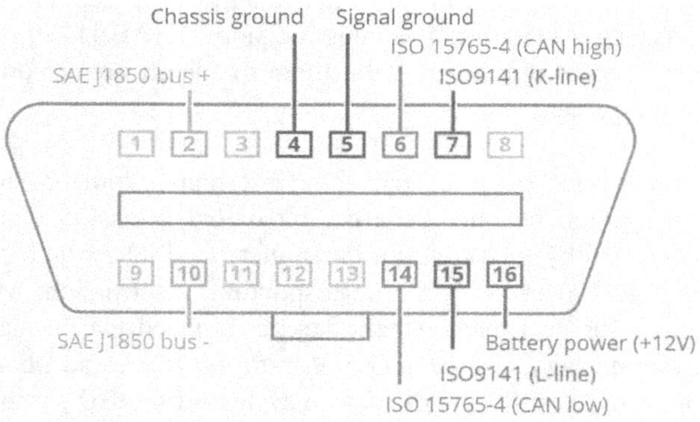

FIGURE 6.2 Female OBD-II Type A connector pinout.

As shown in the figure, SAE J1962 provides for a total of five different diagnostic protocols, leaving the OEMs complete freedom to decide which ones to implement in their vehicles. All diagnostic protocols operate on a request-response communication model, with the diagnostic system requesting information on the status of an electronic component and the diagnostic ECU providing it. Each protocol, however, is based on its own physical line and a different transport mechanism; therefore, not only are they accessible through a different set of pins, but they also differ in their signal levels, data rates, message lengths, and arbitration schemes. For a more detailed discussion, the reader is referred here. Needless to say, each manufacturer has its own preferences when it comes to the diagnostic protocols. It is also worth noting that there is no established rule for determining a priori which protocol is used on a given vehicle; surprisingly, the simplest strategy remains to deduce this information from the pins of its connector. However, this heterogeneity has started to disappear as of 2008, when ISO 15765 has become mandatory for all cars to be marketed in North America; to date, ISO 15765 is the most prevalent vehicle diagnostic protocol around the globe. This growing popularity is also due to the fact that the CAN bus, upon which the protocol relies, is typically already fitted in recent vehicles, which eliminates the need to install an additional physical cable for diagnostic purposes only. This gradual transition is summarised in Tables 6.1 and 6.2. The list of the protocols employed by the various manufacturers for their models can instead be found here.

Furthermore, the predominance of ISO 15765 does not seem destined to disappear any time soon, despite the fact that some OEMs have already embarked on the path of an Ethernet-based diagnostics. The rationale for this is that the diagnostic line is often used to transmit data to the ECUs via Over-the-Air (OTA) updates; Ethernet can then guarantee a considerable reduction in the time that is required to perform these operations, as it provides significantly higher bandwidth than CAN. However, the buzz that Ethernet has brought to the automotive world over the past decade has quickly cooled following the introduction of CAN FD, an evolution of CAN that provides up to a 5-fold increase in data transfer speed. Yet, CAN FD's edge lies in the fact that, unlike Ethernet, it does not require any changes in terms of the physical line. Therefore, any migration to Ethernet will only be evolutionary and not revolutionary.

TABLE 6.1
Diagnostic Protocol History by OEM (Audi-Honda)

Make – Model	1996	1997	1998	1999	2000	2001	2002	2003	2004	2005	2006	2007	2008
Audi/VW	ISO 9141	ISO 9141	ISO 9141	ISO 9141	ISO 9141	ISO 9141	ISO9141 KWP2000	ISO9141 KWP2000	ISO9141 KWP2000	ISO9141 KWP2000 CAN	ISO9141 KWP2000 CAN	ISO9141 KWP2000 CAN	CAN
Bentley	ISO 9141	ISO 9141	ISO 9141	ISO 9141	ISO 9141	ISO 9141	ISO 9141	ISO 9141	ISO 9141 (50%) KWP2000 (50%)	ISO 9141 (35%) KWP2000 (65%)	ISO 9141 (20%) KWP2000 (80%)	CAN	CAN
BMW – Mini	N/A	N/A	N/A	N/A	N/A	N/A	KWP2000	KWP2000	KWP2000	KWP2000	N/A	N/A	CAN
BMW 3-Series, 5-Series, X5, Z3, Z4, Z8, 740i, 740iL, 750iL	ISO 9141	ISO 9141	ISO 9141	ISO 9141	ISO 9141	ISO 9141	ISO 9141	ISO 9141	ISO 9141	ISO 9141	N/A	N/A	CAN
BMW 5-Series; next generation (E60)	N/A	N/A	N/A	N/A	N/A	N/A	N/A	N/A	KWP2000	KWP2000	N/A	N/A	CAN
BMW 745i, 745Li, 765Li	N/A	N/A	N/A	N/A	N/A	N/A	KWP2000	KWP2000	KWP2000	KWP2000			
Chrysler	ISO 9141 (100%)	ISO 9141 (100%)	ISO 9141 (95%) J1850-10.4 (5%)	ISO 9141 (85%) J1850-10.4 (15%)	ISO 9141 (75%) J1850-10.4 (25%)	ISO 9141 (35%) J1850-10.4 (65%)	ISO 9141 (15%) J1850-10.4 (85%)	ISO 9141 (5%) J1850-10.4 (95%)	CAN (5%) J1850-10.4 (95%)	CAN (15%) J1850-10.4 (85%)	CAN (35%) J1850-10.4 (65%)	CAN (85%) J1850-10.4 (15%)	CAN
Daewoo-Lanos	N/A	N/A	J1850-10.4	J1850-10.4	J1850-10.4	J1850-10.4	J1850-10.4	N/A					
Daewoo-Leganza	N/A	N/A	J1850-10.4	J1850-10.4	J1850-10.4	J1850-10.4	KWP2000	KWP2000					
Daewoo-Nubira	N/A	N/A	J1850-10.4	J1850-10.4	J1850-10.4	J1850-10.4	KWP2000	KWP2000					
Ford (w/o exceptions)	J1850-41.6	J1850-41.6	J1850-41.6	J1850-41.6	J1850-41.6	J1850-41.6	5% CAN	5% CAN	50% CAN	85% CAN	90% CAN	CAN	CAN

Exceptions:													
Aspire	ISO 9141	ISO 9141											
Escort 1.8L	ISO 9141												
Probe 2.5L	ISO 9141	ISO 9141											
Villager	ISO 9141	ISO 9141	ISO 9141	ISO 9141	ISO 9141	ISO 9141	ISO 9141						
GM (w/o exceptions)	J1850-10.4	J1850-10.4	J1850-10.4	J1850-10.4	J1850-10.4	J1850-10.4	J1850-10.4	J1850-10.4	80% J1850-10.4 15% CAN 5% KWP2000	55% J1850-10.4 40% CAN 5% KWP2000	25% J1850-10.4 70% CAN 5% KWP2000	25% J1850-10.4 70% CAN 5% KWP2000	CAN
Exceptions:													
Saturn LS 3.0L	N/A	N/A	N/A	N/A	KWP2000	KWP2000	KWP2000	KWP2000					
Saturn VUE 3.0L	N/A	N/A	N/A	N/A	N/A	N/A	KWP2000	KWP2000					
Saturn VUE 3.5L	N/A	N/A	N/A	N/A	N/A	N/A	N/A	KWP2000					
Saturn ION	N/A	N/A	N/A	N/A	N/A	N/A	N/A	CAN					
Cadillac Catera	N/A	KWP2000	KWP2000	KWP2000	KWP2000	KWP2000	N/A	N/A					
Cadillac CTS	N/A	N/A	N/A	N/A	N/A	N/A	N/A	KWP2000					
GEO Metro	ISO 9141	ISO 9141	ISO 9141	ISO 9141	ISO 9141	ISO 9141	N/A	N/A					
GEO Prism	J1850-10.4	J1850-10.4	ISO 9141	ISO 9141	ISO 9141	ISO 9141	ISO 9141	N/A					
GEO Tracker	ISO 9141	ISO 9141	ISO 9141	KWP2000	KWP2000	KWP2000	KWP2000	KWP2000					
Pontaic Vibe	N/A	N/A	N/A	N/A	N/A	N/A	N/A	ISO 9141					
Saab 9-3	N/A	N/A	N/A	N/A	N/A	N/A	N/A	CAN					
Honda/Acura	ISO 9141	ISO 9141	ISO 9141	ISO 9141	ISO 9141	ISO 9141	ISO 9141	ISO 9141	ISO 9141	ISO 9141	70% ISO 9141 30% CAN	40% ISO 9141 60% CAN	CAN

TABLE 6.2
Diagnostic Protocol History by OEM (Hyundai-Volvo)

Make – Model	1996	1997	1998	1999	2000	2001	2002	2003	2004	2005	2006	2007	2008
Hyundai	ISO 9141	ISO 9141	ISO 9141	KWP2000	ISO 9141 KWP2000	ISO 9141 KWP2000	ISO 9141 KWP2000	ISO 9141 KWP2000	ISO 9141 KWP2000	ISO 9141 KWP2000	ISO 9141 KWP2000	ISO 9141 KWP2000	CAN
Isuzu	J1850-10.4	J1850-10.4	J1850-10.4	J1850-10.4	J1850-10.4	J1850-10.4	J1850-10.4	J1850-10.4	J1850-10.4	J1850-10.4	N/A	N/A	CAN
Jaguar	ISO 9141	ISO 9141	ISO 9141	ISO 9141	ISO 9141	ISO 9141	ISO 9141	ISO 9141	ISO 9141	ISO 9141	ISO 9141		
Kia – Optima	N/A	N/A	N/A	N/A	N/A	KWP 2000	KWP 2000	KWP 2000	KWP 2000	KWP 2000			
Kia – Rio	N/A	N/A	N/A	N/A	N/A	ISO 9141	ISO 9141	ISO 9141	ISO 9141	ISO 9141			
Kia – Sephia	ISO 9141	ISO 9141	ISO 9141	ISO 9141	ISO 9141	ISO 9141	N/A	N/A	N/A	N/A	N/A	N/A	CAN
Kia – Spectra	N/A	N/A	N/A	N/A	ISO 9141	ISO 9141	ISO 9141	ISO 9141	ISO 9141	ISO 9141			
Kia Sportage	ISO 9141	ISO 9141	ISO 9141	ISO 9141	ISO 9141	ISO 9141	ISO 9141	ISO 9141	ISO 9141	ISO 9141			
Kia Sedona	N/A	N/A	N/A	N/A	N/A	N/A	ISO 9141	ISO 9141	ISO 9141	ISO 9141			
Kia Sorento	N/A	N/A	N/A	N/A	N/A	N/A	N/A	KWP 2000	KWP 2000	KWP 2000			
Land Rover	ISO 9141	ISO 9141	ISO 9141	ISO 9141	ISO 9141	ISO 9141	0% CAN	0% CAN	0% CAN	62% CAN	CAN	CAN	CAN
Mazda	ISO 9141	ISO 9141	ISO 9141	ISO 9141	ISO 9141	ISO 9141	ISO 9141	75%-ISO 9141 25%-CAN	60%-CAN 40%-ISO 9141	60%-CAN 40%-ISO 9141	CAN	CAN	CAN
Mazda-Ford	J1850-41.6	J1850-41.6	J1850-41.6	J1850-41.6	J1850-41.6	J1850-41.6	J1850-41.6	J1850-41.6	J1850-41.6	50%-J 1850 50%-CAN	50%-J 1850 50%-CAN	Tribute-CAN Truck?	CAN
Mercedes	ISO 9141	ISO 9141	ISO 9141	ISO 9141	ISO 9141 KWP2000	ISO 9141 KWP2000	ISO 9141 KWP2000	KWP2000	KYP2000	KYP2000 CAN	KYP2000 CAN	KYP2000 CAN	CAN
Mitsubishi	ISO 9141	ISO 9141	ISO 9141	ISO 9141	ISO 9141	ISO 9141	ISO 9141	ISO 9141	ISO 9141	ISO 9141	ISO 9141 CAN	N/A	CAN
Nissan/Infiniti	ISO 9141	ISO 9141	ISO 9141	ISO 9141	ISO 9141	ISO 9141	ISO 9141	ISO 9141	ISO 9141	ISO 9141	ISO 9141	N/A	CAN
Porsche	ISO 9141	ISO 9141	ISO 9141	ISO 9141	ISO 9141	ISO 9141	ISO 9141	ISO 9141					
Rolls Royce	ISO 9141	ISO 9141	ISO 9141	ISO 9141	ISO 9141	ISO 9141	ISO 9141	ISO 9141	KWP 2000	KWP 2000	KWP 2000	N/A	CAN
Subaru	ISO 9141	ISO 9141	ISO 9141	ISO 9141	ISO 9141	ISO 9141	ISO 9141	ISO 9141 KWP 2000	ISO 9141 KWP 2000	KWP 2000	CAN	CAN	CAN

Suzuki Aerio 2.0L	N/A	N/A	N/A	N/A	N/A	N/A	KWP 2000	KWP 2000	KWP 2000	KWP 2000	KWP 2000	CAN	CAN
Suzuki Esteem 1.6L	ISO 9141	ISO 9141	ISO 9141	ISO 9141	ISO 9141	ISO 9141	ISO 9141	N/A	N/A	N/A	N/A	N/A	CAN
Suzuki Esteem 1.8L	N/A	N/A	N/A	ISO 9141	ISO 9141	ISO 9141	ISO 9141	N/A	N/A	N/A	N/A	N/A	CAN
Suzuki Grand Vitara 2.5L	N/A	N/A	N/A	ISO 9141	ISO 9141	ISO 9141	ISO 9141	KWP 2000	KWP 2000	KWP 2000	KWP 2000	CAN	CAN
Suzuki Grand Vitara 2.7L	N/A	N/A	N/A	N/A	N/A	KWP 2000	KWP 2000	KWP 2000	KWP 2000	KWP 2000	KWP 2000	CAN	CAN
Suzuki Sidekick 1.6L	ISO 9141	ISO 9141	ISO 9141	N/A	N/A	N/A	N/A	N/A	N/A	N/A	N/A	N/A	CAN
Suzuki Sidekick 1.8L	ISO 9141	ISO 9141	ISO 9141	N/A	N/A	N/A	N/A	N/A	N/A	N/A	N/A	N/A	CAN
Suzuki Swift 1.3L	ISO 9141	ISO 9141	ISO 9141	ISO 9141	ISO 9141	ISO 9141	N/A	N/A	N/A	N/A	N/A	N/A	CAN
Suzuki Vitara 1.6L	N/A	N/A	N/A	ISO 9141	ISO 9141	ISO 9141	N/A	N/A	N/A	N/A	N/A	N/A	CAN
Suzuki Vitara 2.0L	N/A	N/A	N/A	ISO 9141	ISO 9141	ISO 9141	ISO 9141	ISO 9141	ISO 9141	ISO 9141	ISO 9141	CAN	CAN
Suzuki X90 1.6L	ISO 9141	ISO 9141	ISO 9141	N/A	N/A	N/A	N/A	N/A	N/A	N/A	N/A	N/A	CAN
Toyota/Lexus/ Scion	J1850-10.4	J1850-10.4 ISO 9141	J1850-10.4 ISO 9141	J1850-10.4 ISO 9141	ISO 9141	ISO 9141	ISO 9141	ISO 9141	ISO 9141 CAN	ISO 9141 CAN	ISO 9141 CAN	N/A	CAN
Volvo	ISO 9141	ISO 9141	ISO 9141	ISO 9141	ISO 9141	ISO 9141	ISO 9141	ISO 9141	90% ISO 9141 10% CAN	5% ISO 9141 95% CAN	CAN	CAN	CAN

The high prevalence of CAN bus in the automotive market offers a further alternative for accessing vehicle data: *sniffing the messages* transiting the diagnostic network, which is practicable on virtually every vehicle. CAN bus – the physical layer on which the diagnostic protocol is built – supports in fact broadcast data transmission, permitting the capture of any frames sent to any ECU. This means that a dongle, if equipped with a CAN interface, can receive all messages exchanged on the vehicle's diagnostic network without resorting to the diagnostic procedure; since the diagnostic network typically carries engine data, the message-sniffing approach would enable the OBD dongle to access much of the data required for fleet management. This strategy would even guarantee a relatively high update rate: the repetition time of the messages circulating the CAN bus is estimated to be in the range of 10-50 milliseconds, which is significantly shorter than the data collection time achievable with the request-response approach of ISO 15765, which can hardly fall below one second. Despite this, the recourse to the diagnostic protocol is still to be preferred. Indeed, apart from the fact that a data acquisition frequency higher than 1 Hertz proves to be excessive for any fleet management application, the dataset that is obtainable via message sniffing turns out to be not immediately usable. In fact, although the payload of the messages travelling on CAN is not commonly encrypted, the message content is made inaccessible through a strategy known as *security through obscurity*, which consists of encoding the data using a formula known only to the manufacturer. Access to this dataset then requires either the recourse to the OEM's coding scheme, which is usually unavailable, or to reverse engineering, in the attempt to fathom these formulas through deductive reasoning. It goes without saying that the latter strategy is not only exceptionally complex, but also extremely impractical, as the coding scheme may even vary between different models of the same manufacturer.

6.1.3 Alternatives to the OBD Dongle

Today, there is an alternative for accessing vehicle diagnostic data that does not require the use of a dongle. In apparent self-contradiction, this strategy has been promoted by the manufacturers themselves, who, unable to inhibit the use of the OBD suite for service development, have managed to turn a disadvantageous situation into a valuable business opportunity. In fact, incapable of preventing monetisation to third parties, the OEMs have chosen to compete with them by handing over their vehicle data to new market players, like Otonomo, Mojio, and Zubie, who can redistribute it for service-development purposes. All these companies make use of a service platform that ingests, cleans, aggregates, and secures data, and delivers application-ready, enriched datasets and insights. Not only does this approach eliminate a significant portion of the effort required to launch services based on vehicle data, but it also removes the need for an OBD dongle. Nowadays, in fact, an increasing number of cars feature built-in connectivity capabilities, both to obtain OTA updates and to transmit the vehicle data to the manufacturer for internal use. A win-win situation for all but the dongle providers (Figure 6.3).

FIGURE 6.3 Otonomo value proposition.

Therefore, although not yet reported in any studies, a more convenient alternative to using an OBD dongle is available to access diagnostic data. Relying on an automotive data service provider proves extremely advantageous for the development of services based on aggregated data, such as congestion management, traffic flow, and route optimisation, transport system planning, and parking management. This approach lends itself particularly well also to the development of passenger vehicle services, allowing owners to monitor and track the health of their cars, or intervene promptly in the event of theft thanks to geofencing alerts, all without the need to purchase and rely on a third-party OBD device of dubious origin. We believe, however, that this solution is not as beneficial for fleet management applications. Indeed, it should be noted that a generic fleet, unless it is managed by the manufacturer itself – as in the case of the Drive Now car-sharing service by BMW – is composed of vehicles from different OEMs. Therefore, the more heterogeneous the fleet, the greater the risk of not being able to access all necessary data, as some manufacturers may not be covered by the chosen data service provider. Moreover, fleet vehicles may lack built-in connectivity, making their data inaccessible through the data provider. Finally, with this approach, the vehicles would serve solely as a data source, making it impossible to establish two-way communication with them; neither would it be feasible to carry out on-board processing, nor to control essential parameters such as data collection or transmission frequencies.

In light of the findings of this section, the use of an OBD dongle that supports the ISO 15765 diagnostic protocol represents the most suitable strategy to access the vehicle data required by a context-aware, real-time fleet management application. However, as will be evidenced in the rest of this chapter, it still remains necessary to design both the hardware and software of the dongle so as to ensure that such a vehicle interface can meet the system requirements of the target application.

6.2 DESIGNING THE VEHICLE INTERFACE

Over the last few years, OBD-II dongles have flooded the automotive aftermarket. Such devices, often referred to as OBD readers, allow anyone, for a modest outlay, to monitor and keep track of their vehicle's speed, distance travelled, and fuel consumption, as well as obtain detailed information on a variety of car malfunctions. Accessing this information typically requires a mobile phone application, which receives the data gathered by the dongle via a Bluetooth interface. Evidently, these devices are specifically designed for the passenger vehicle market and, as such, do not lend themselves that well to fleet management applications; fleet operations, in fact, require OBD dongles with fundamentally different capabilities, such as the ability to transmit car data to remote servers, where it can be both processed and stored. The obvious solution is to rely on companies like Geotab, Webfleet, and Samsara, which not only provide a more suitable OBD-II dongle but also the infrastructure to transmit, process, store, and retrieve the diagnostic data obtained from the fleet. However, as has been reiterated several times throughout this book, these solutions are designed for vehicles with radically different requirements than those of an emergency fleet, making their use extremely problematic. In fact, the CVPs these companies provide are intended for traditional fleet management, which, as already explained, does not require high data collection or transmission frequencies; needless to say, these parameters cannot be adjusted at will. Neither is any form of on-board processing allowed, since the OBD-II dongles act as mere data transmitters, nor is it possible to modify the communication protocol the diagnostic data is transmitted with, which is typically HTTP. Therefore, as context-aware, real-time fleet management applications require the vehicle data to be handled in a way that is not yet mainstream in today's market, the only feasible solution is to build an OBD dongle from the ground up.

To build the OBD dongle, we could therefore consider leveraging the knowledge available in the academic literature, as this topic is far from unexplored in research. H.-S. Kim, S.-J. Jang and J.-W. Jang, for example, develop an OBD scanner that shares vehicle engine fault data over Bluetooth with an Android application. The work of C. E. Lin et al. leads, instead, to a dongle that is built around an ELM327 microcontroller for the monitoring of vehicle pollution, which transmits the diagnostic information to a mobile phone application via the RS-232 interface. Y. Yang develops a device to acquire real-time diagnostic data from hybrid-electric vehicles, while J. E. Meseguer designs an OBD dongle to quantify the impact of driving behaviour on fuel consumption; both studies feature a phone application to analyse and visualise the diagnostic information obtained from the car. It would be easy to fill up entire pages with examples of research projects that led to the development of an OBD dongle; yet none of these devices would prove capable of meeting the unique requirements of context-aware, real-time fleet management applications.

In light of the above, to ensure the effective management of the ambulance fleet of the NHS, it is imperative to depart from both the OBD-II dongles currently available on the market and the various solutions presented in the academic literature.

Identifying the data source to be used, its physical access point, and the method of obtaining this data, as done in Section 6.1, proves therefore insufficient. Indeed, it is of equal importance to design and build the dongle in such a way that its behaviour can meet the specific requirements of the applications it is intended to support. Starting from the behaviour expected from the OBD device, outlined in Section 6.2.1, Section 6.2.2 will describe the type of hardware to be used for its implementation, while Section 6.2.3 will discuss both the structure and functions of its software.

6.2.1 Purpose and Intended Use

The vehicle interface is the component of the CVP responsible for collecting diagnostic data from the vehicle, contextualising it at the vehicle level, and communicating it to the system backend. Following on from Section 6.1, this component comes in the form of an OBD-II dongle that supports ISO 15765 as the only diagnostic protocol. Being connected to the OBD port, this device can directly be powered from the vehicle battery, which eliminates the need for a dedicated power supply. In spite of this, it is still advisable to provide the dongle with an additional power source in order to preserve its operations in the unlikely event of a loss of power due to a vehicle malfunction. In the case of the ambulance fleet, the dongle can leverage the auxiliary battery used to power the on-board medical equipment.

Our dongle is designed to be extremely user-friendly so that it can be installed and operated without any particular expertise. For this very reason, it is provided with an automatic *boot* function, which is executed as soon as this device is connected to the diagnostic port. It is essential that the dongle is only connected to the vehicle when the car engine is switched off, so as to avoid introducing a new node into the diagnostic network when the latter is already in operation. In addition, this allows the dongle to initialise its navigation module when there are no measurements to be acquired yet, which is of utmost importance as this module requires an interval of about 30–45 seconds – known as *cold start* – to start providing valid position data. Therefore, if a shorter interval elapses between dongle boot-up and vehicle engine start-up, the dongle may share a few measurements with invalid GPS data. To overcome this, our device replaces invalid position data with a default value, which can then be rectified by a GPS correction algorithm before these measurements are stored in the database. The same occurs in the case of loss of GPS signal.

The actual configuration of the device, however, can only occur when the car engine is running, as this is when messages are permitted on the diagnostic network. The *Set-up* operation, which is automatically performed upon engine ignition, allows the dongle to determine whether the ambulance it is connected to is compatible with the ISO 15765 diagnostic protocol. If compatibility is confirmed, the device can configure itself according to the type of diagnostic communication supported and retrieve the list of parameters obtainable from the diagnostic ECU. Should this be the first configuration of the device, the set-up leads to the creation of a dongle instance in the database; otherwise, the set-up merely retrieves the dongle ID from the storage unit. During the set-up operation, the device also allows for the configuration of the vehicle it is connected to, following an approach similar to the one described above. The only difference is that, in addition to the vehicle ID, the dongle also retrieves the parameters of the emission model that is associated with the vehicle. As explained in Section 5.2.2, this information is exchanged via HTTP.

If the configuration is successful, the OBD dongle can proceed with the *extraction of diagnostic data* via the ISO 15765 protocol. The management of the ambulance fleet of the NHS, at least in the current version of the system, requires two vehicle data items only, vehicle speed and engine speed, which can be retrieved using the aforementioned protocol via Mode 01. In light of the fact that (NFR3) the CVP should acquire and store the data required for processing as soon as it becomes available, this device uses a data collection frequency of 1 Hertz: besides being impractical, in fact, a higher value would not provide any substantial benefit to the application. Before being communicated, the diagnostic data is used by the dongle to estimate the emissions produced by the ambulance. This is made possible by the high computing capability of the device – not found in a conventional OBD reader – which is indispensable for implementing security mechanisms at the dongle level. Moreover, delocalising emission estimation on the dongle, through distributed processing, provides significant advantages in terms of database performance. In fact, since each ambulance is characterised by its own emission model, if the estimation of its pollutants were to be performed in the system backend, each measurement would require two database queries: one to retrieve the emission model associated with that vehicle, and the other to store the obtained estimate within the database. Therefore, the on-board processing halves the number of database queries, as the emission model of a vehicle is only requested once during the set-up of its dongle. Diagnostic data, once enriched with emission estimates, is locally contextualised: this task provides each measurement with a timestamp, the GPS position where it was recorded, the identifiers of the vehicle it was obtained from, and the identifier of the dongle responsible for its transmission. Finally, each measurement is shared via MQTT, using the best-performing cellular network technology available to the dongle's communication module at the time of transmission.

When the vehicle engine is switched off, the dongle can interrupt the collection of diagnostic data, as the car's sensors stop generating any information; apart from this, the device must keep all its functionalities active. This is in stark contrast to the behaviour of the various OBD devices that are available on the market, such as those used for pay-as-you-drive (PAYD) services, which can enter a sleep mode until the next engine start. For emergency fleet management, instead, the device must keep collecting and transmitting information on vehicle location even when the engine is not running, as this data is critical for decision-making. In this scenario, it is essential to minimise the power consumption of the dongle, since the device could cause parasitic drainage of the battery, an eventuality that is clearly to be avoided. The very moment the device is disconnected from the power supply, all operations are interrupted. In view of the above, the dongle's behaviour can be modelled as per the state machine diagram presented in Figure 6.4. The sequence diagrams of Figures 6.5 and 6.6 summarise, instead, its interactions with the other components of the CVP while performing configuration and data sharing tasks.

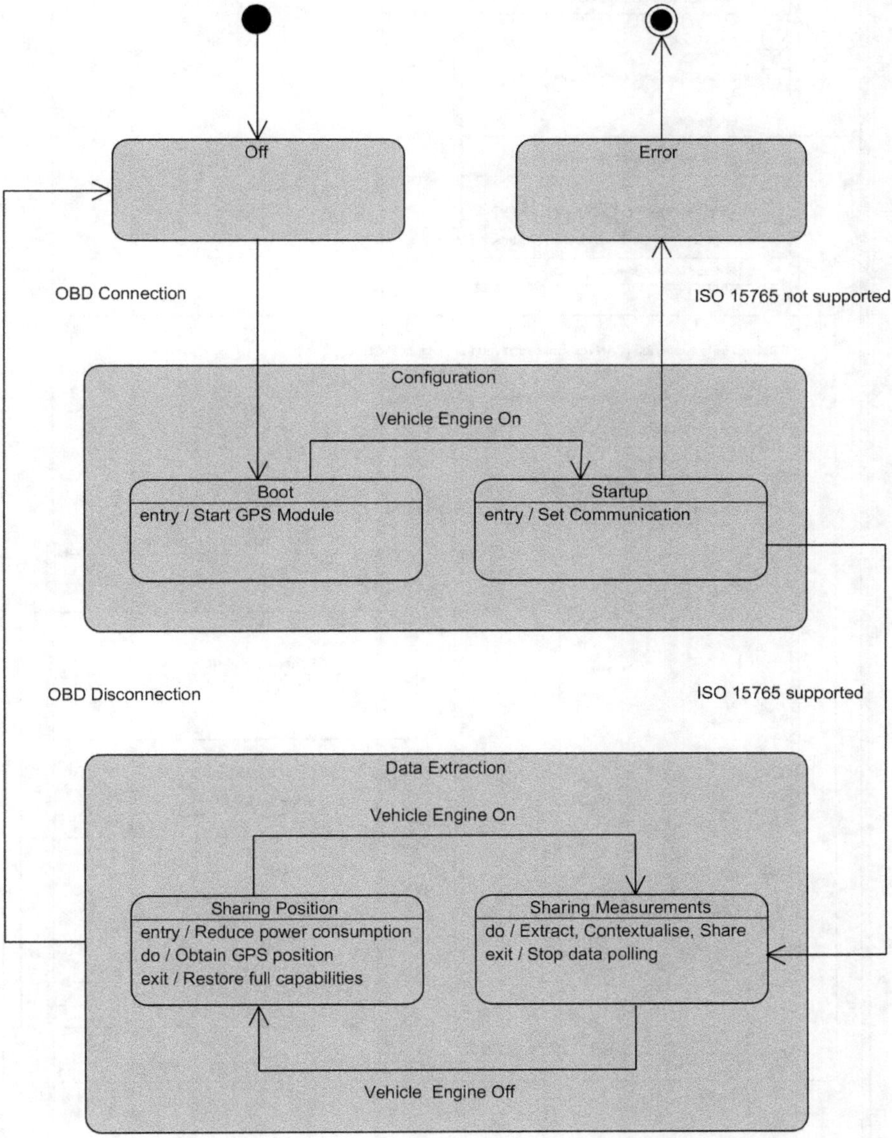

FIGURE 6.4 OBD dongle: State machine diagram.

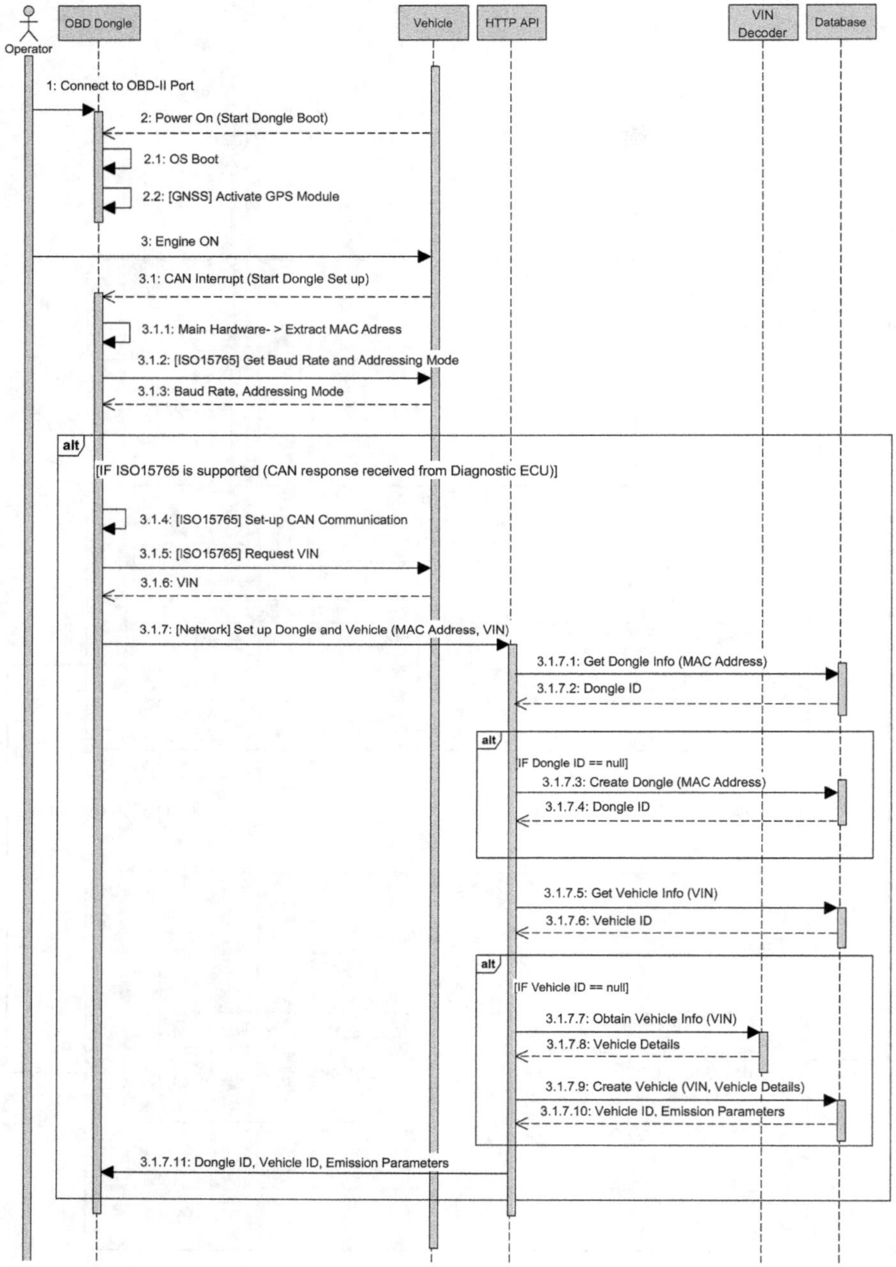

FIGURE 6.5 OBD dongle configuration: Sequence diagram.

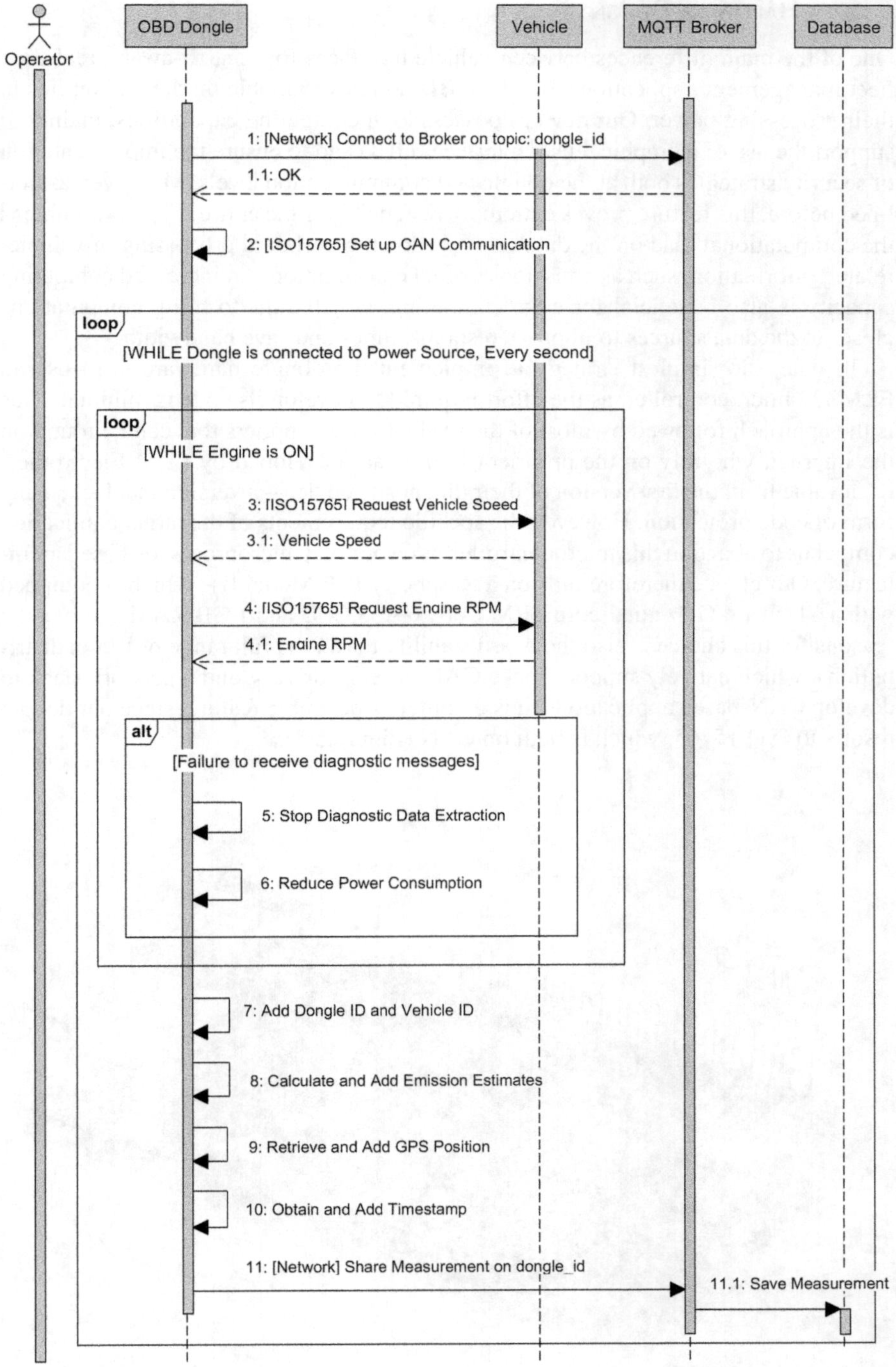

FIGURE 6.6 OBD dongle data extraction: Sequence diagram.

6.2.2 Hardware Design

One of the main differences between vehicle interfaces for context-aware, real-time fleet management applications and the OBD dongles available on the market lies in their processing power. Our devices possess high computing capabilities, mainly to support the use of a graphical user interface (GUI) and to ensure the implementation of security strategies both at the dongle and communication levels. Moreover, as outlined before, this feature proves extremely beneficial for the entire CVP, as it lightens the computational load on the database by allowing on-board processing of vehicle-related information, such as emissions or fuel consumption. An increased computing capacity is also beneficial for edge computing, which aims to bring computations closer to the data sources to improve response times and save bandwidth.

To date, the simplest strategy to implement the dongle hardware is to use an ELM327 microcontroller, as the effort required to develop its code is minimal. This is the approach followed by most of the OBD dongle suppliers that can be found on the Internet, who rely on the proprietary firmware developed by ELM Electronics, retrievable from the first version of their diagnostic device – a version that lacked any form of code protection. However, the specific requirements of the target application compel us to abandon this microcontroller in favour of a microprocessor-based architecture. Our choice therefore falls on a Raspberry Pi 3 Model B+, which is equipped with a 64-bit 1.4 GHz quad-core ARM Cortex-A53 CPU and 1 GB RAM. Among the reasons for this choice is also the board's ability to run the full range of Linux distributions, which natively support SocketCAN, a set of drivers, and a network stack to develop CAN-based applications; it is definitely a desirable feature, since our dongle resorts to ISO 15765, which is built on CAN (Figure 6.7).

FIGURE 6.7 Raspberry Pi 3 Model B+.

The communication between dongle and vehicle is enabled by the PiCAN2 board, which provides the Raspberry Pi with CAN bus capabilities. It consists of an MCP2515 CAN controller with a MCP2551 CAN transceiver, a DB9 connector, which allows the dongle to be plugged into the vehicle via an OBDII to DB9 adapter cable, and an SMPS module, to power the device directly from the diagnostic port. The version of the board we have opted for, however, does not include sensors like accelerometers and gyroscopes, superfluous in the current platform configuration. Among the features of the board, it is worth mentioning the use of the GPIO25 pin for receiving interrupts, signals that request the CPU of Raspberry Pi to suspend its current execution and service a user-defined routine. This is a mechanism that proves invaluable to fleet management: in fact, when the GPIO25 pin is connected to one of the two CAN wires accessible from the OBD socket, the dongle becomes capable of detecting changes in the status of the diagnostic network. Interrupts can then be used to initiate dongle setup upon engine ignition, without the burden of a CAN bus polling activity. Failure to receive messages, due to engine shutdown, prompts the dongle to terminate the acquisition of the diagnostic data (Figure 6.8).

FIGURE 6.8 PiCAN2 CAN interface board with SMPS.

Neither the Raspberry Pi nor the PiCAN2 board features built-in communication or GNSS positioning capabilities. It is therefore essential to equip the dongle with a further add-on that incorporates two modules, one dedicated to navigation and positioning, and the other to cellular network communication. Our preference is for the Waveshare SIM7600E-H, a 4G dongle that connects with the main board via USB. To design a diagnostic device with no external attachments, however, we would instead need to use the SIM7600E-H 4G HAT, which is equivalent to the latter in terms of functionalities and performance. When using this component, it is indispensable to ensure that there is no conflict between the two hats in terms of GPIO usage. Both Waveshare devices allow the use of a wide range of communication protocols – including HTTPS and MQTT – and communication technologies, like LTE Cat-4, LTE-TDD, 3G, and 2G. In terms of GNSS positioning, they support technologies such as GPS, GALILEO, Glonass, BeiDou, QZSS and LBS base station positioning. Being to all intents and purposes modems – namely devices that convert data from the digital format to one suitable for transmission on analogue media and vice versa – these devices require the use of the Hayes command set, also known as AT commands, to be operated (Figure 6.9).

(a) (b)

FIGURE 6.9 Waveshare modules. (a) SIM7600E-H 4G dongle. (b) SIM7600E-H 4G hat.

Finally, ambulance operators are expected to use this device to access key data, such as the ID of the vehicle to which the dongle is connected – information they use to start a shift via their phone app – as well as the emergency route to be followed after the vehicle has been dispatched. Hence, the decision to use a 7-inch touchscreen display with support for ten-finger touch and an on-screen keyboard, which eliminates the need for a physical keyboard and mouse. The connection with the Raspberry Pi, as shown in Figure 6.10, is established via a ribbon cable, while power is supplied through the GPIO port.

FIGURE 6.10 Raspberry Pi touch display.

Based on the above considerations, we developed the OBD dongle prototype shown in Figure 6.11, equipped with a 3D-printed casing and named VEICOLNET – a code name that is also be used to identify the user interfaces of the data platform. Although ideal for testing purposes, this device proves unsuitable for commercial uses. The recourse to off-the-shelf components has resulted in an over-designed device, meaning that the OBD dongle is equipped with components that do not bring any tangible benefits to the system. Among these are the Ethernet, Bluetooth and Wi-Fi modules of the Raspberry Pi, its USB and HDMI ports, and the AUX connector of the Waveshare board. This results in high power consumption, which can, however, be limited by disabling unnecessary device components via software. Still, the issue of the device's high production cost would remain. If the commercialisation of the dongle were to be considered, we therefore recommend designing the hardware of the device from the ground up, incorporating only those components that are strictly necessary to support the target application.

FIGURE 6.11 OBD-II dongle prototype VEICOLNET.

6.2.3 Software Design

The software behind the operations of the dongle is built around Node.js, an open-source, cross-platform JavaScript runtime environment that has already been used for the implementation of the API of our platform, as detailed in Section 5.2.2. Node.js is an exceptionally powerful tool not only for server-side development, but also for that on the client side. This is particularly true for IoT applications, given that Node.js natively supports the main protocols used in this domain, such as HTTP and MQTT. Considering that our dongles rely on both, our decision could not help but lean towards this technology. Not to be underestimated is also the fact that it provides a number of mechanisms for authenticating users and devices and protecting communicated data. In addition, Node.js compiles JavaScript into native machine code, which guarantees high execution speed, making it the perfect solution for highly responsive applications. This category clearly includes context-aware, real-time fleet management applications, for which it is essential (NFR3) that dongles acquire and share data as soon as it becomes available. Due to its shorter execution time, an application built on Node.js also uses fewer resources to run than those based, for example, on Java or Ruby. Admittedly, this is an invaluable feature for ordinary IoT applications, but somewhat less so for fleet management, as dongles are not battery-powered but connected to the vehicle's electrical systems.

The software of our dongle has a modular structure, consisting of a main, which starts at boot, and two independent JavaScript files dedicated to the core functionalities of the device – its configuration and the sharing of measurements; the former relies on Axios to share the dongle configuration data with the API, while the latter uses MQTT to forward the measurements to the broker. Both files make use of native add-ons – dynamically-linked shared objects written in C/C++ that can be loaded and used as ordinary Node.js modules. These add-ons represent the real engine of the software, as they enable control of the hardware, which would not be possible with the resources provided by the JavaScript library alone. As shown in Figure 6.12, the layout of the software mirrors the hardware structure of the dongle, being based on three sets of add-ons, one for each device component. One of these add-ons allows the dongle to retrieve diagnostics data from the vehicle via the PiCAN2 board, another enables the execution of the AT commands through the Waveshare dongle – either to check the status of the cellular network or to obtain GNSS information; a final add-on provides access to the MAC address of the main board. An in-depth description of these modules can be found on the following pages, while their code is in Appendix D.

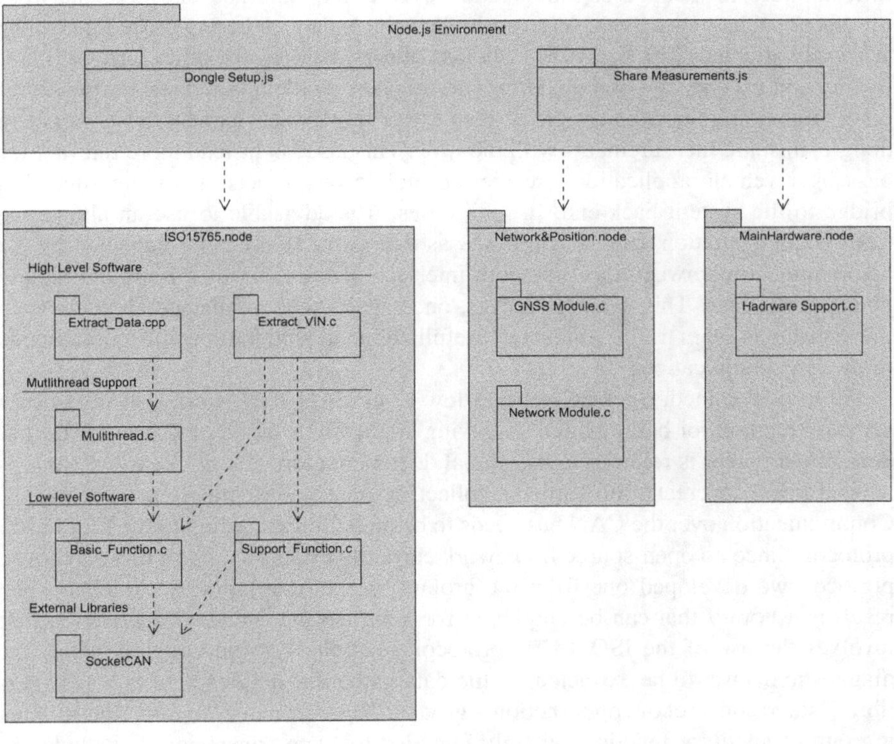

FIGURE 6.12 OBD dongle software: Package diagram.

6.2.3.1 ISO 15765 Diagnostic Software

The add-on responsible for obtaining data from the vehicle is based on SocketCAN, an implementation of CAN protocols for Linux operating systems. SocketCAN uses the Berkeley socket API, integrates with the Linux network stack, and implements CAN device drivers as network interfaces. Its use results in highly portable software, which is highly beneficial in scenarios where hardware needs to be replaced often. The software obtained with SocketCAN is also considerably more sophisticated than conventional implementations of CAN protocols, which are typically distributed as device drivers for specific hardware. Such device-specific software not only lacks generality, being hardware-dependent, but also provides limited functionality, as it is designed for character devices and thus based on the transmission and reception of single bytes. This limitation has a second-order effect: transport-layer protocols and mechanisms such as frame queuing must be implemented within the user-space application, and only a single open process can be used at a time, similar to serial interfaces. Another major advantage of SocketCAN is its support for CAN FD, which makes it particularly suitable for fleet management applications, as it ensures software usability even in the event of a technological transition of the vehicle diagnostic protocol to this newer version of CAN. However, SocketCAN may prove unsuitable for real-time or highly reliable applications, as it relies on the same data handling infrastructure to manage communication over CAN, Ethernet, and other network interfaces such as Bluetooth. As a result, the vehicle data gathering procedure can be adversely affected, with delays or even loss of messages, by the other forms of communication used by the dongle. However, this is a problem that does not pertain to our applications, since the dongles use an external interface, namely the Wavershare dongle, to share measurements with the CVP. This issue is instead more relevant for passenger vehicle applications, where the mobile phone acts as a communication bridge to the system back-end. In such cases, it is advisable to use an alternative CAN implementation, such as LinCAN, as data extraction can be impacted by the communication between the Bluetooth interface of the main board and the mobile phone of the user. This example proves, once again, that any design choice – even those that may seem trivial – must be carefully aligned with the specific requirements of the target applications.

As just explained, SocketCAN is a low-level CAN framework that provides a standard method for building and deploying applications based on CAN; for further details, the reader is referred to its official documentation. Yet, SocketCAN alone is not sufficient to create software for collecting diagnostic data from the vehicles. Communication over the CAN bus needs to be modelled according to the ISO 15765 protocol. Since no open-source framework currently exists to support this diagnostic protocol, we developed one from the ground up, building upon SocketCAN. The result is a library that can be employed for any fleet management application that involves the use of the ISO 15765 protocol, regardless of the target vehicles, the diagnostic dataset to be extracted, or the data gathering frequency to be used. This library stands on a set of core function – grouped together in *Basic_Function.c* – plus a group of ancillary functions, mainly intended for type conversions – included in *Support_Function.c*. A brief description of the most significant functions follows below:

Set_BaudRate is the function that determines whether the diagnostic network of the vehicle is CAN-based. If so, it returns the communication speed that the dongle must employ to interact with the vehicle's diagnostic ECU, which, according to ISO 15765, can only assume values of 250 and 500 kilobits/s; otherwise, it returns a value indicating that no CAN-based diagnostic protocol can be used. This function does not require the transmission of any CAN messages, as it operates by sniffing the messages that circulate on the diagnostic network, which, if CAN-based, are transmitted via broadcast.

Set_AddressingFormat is the function that determines whether the vehicle supports the ISO 15765 protocol, assuming its diagnostic network is CAN-based. If so, it returns the address the dongle should use to communicate with the vehicle's diagnostic ECU, as well as the address range from which diagnostic data should be received. The former is used to label all messages that the dongle transmits over the CAN bus, while the latter is used to set a software filter for incoming messages, preventing the device from intercepting all communications that take place on the diagnostic network. All the addresses obtained downstream of this function depend on the type of CAN protocol the diagnostic network uses; either 11-bit addresses are possible, if CAN 2.0A is in use, or addresses with 29 bits, typical of CAN 2.0B, which is becoming increasingly prevalent given the growing number of nodes connected to this bus. This function also provides the list of the PIDs that the OBD dongle can extract from the vehicle. All this information is acquired by exchanging ISO 15765 messages: failure to obtain this information implies that the diagnostic network, although CAN-based, does not support the chosen diagnostic protocol.

Send_ISO15765Message is the function to be invoked to request a specific data item from the vehicle's diagnostic ECU via an ISO 15765 message. It takes as input the PID service to be employed, the requested PID, and the list of the PIDs supported by that vehicle, which is used to assess whether the chosen parameter can be obtained or not. It also requires the address that the dongle should adopt to communicate with the diagnostic ECU, which is not necessarily the default one, as some OEMs may use non-standard IDs. It returns the status of the operation.

Receive_ISO15765Message is the function used to fetch an ISO 15765 message from the incoming queue. It relies on a switch statement that decodes the payload using a formula that depends on the PID associated with the message, as per SAE J1979. The function returns the content of the message received on the CAN bus.

All the functions described above, with the exception of the one for determining the baud rate of the diagnostic network, resort to SocketCAN, as they use a socket bound to a CAN interface, either for sending or receiving an ISO 15765 message. These functions are then combined to implement more complex software modules, like *Extract_VIN.c*, which allows the vehicle identification number to be retrieved from the diagnostic ECU. This module is in charge of a significant portion of the dongle setup operations: it first establishes whether the vehicle supports ISO 15765 as a diagnostic protocol, by using *set_BaudRate* and *set_AddressingFormat*, then proceeds with the extraction of the VIN, which requires the transmission and reception of several diagnostic messages. The VIN, due to its size, is in fact split over several CAN messages, in line with the extended addressing strategy set out in 15765-2; this results in a rather articulated message exchange between the dongle and the diagnostic ECU, as evidenced by the code in the appendix. *Extract_VIN.c*, however, is not responsible for decoding the obtained string, a task that is delegated to the VIN Decoder by Vincario, which receives this string via an HTTP request. After all, VIN decoders provide more information about the vehicle than what is actually encoded in the VIN, as these APIs cross-reference data from a variety of sources, such as driver databases, legal registers, and more.

To make use of the functionality of *Extract_VIN.c* in Node, it is imperative to encapsulate its code in a native add-on, as was explained in the previous pages. To this end, we have chosen N-API, a stable and V8-independent API for generating add-ons for Node.js: the result is a native add-on that, unlike those obtainable with NAN, proves to be extremely reusable, since it can be deployed on different versions of Node without the need to recompile its code. *DongleSetup.js* leverages the capabilities of *Extract_VIN.c* synchronously, meaning that once the native add-on is executed, the dongle waits for its completion before retrieving the VIN and resuming its workflow.

The module responsible for retrieving vehicle data from the diagnostic control unit is *Extract_Data.cpp*, which employs a fundamentally different approach from that used by conventional OBD readers. This module leverages multi-threading to introduce parallelism into the data extraction procedure, allocating requests for the various PIDs across multiple *pthreads*. This strategy not only makes data collection more responsive but also extremely scalable, since the software generates the phtread pool depending on the application it has to support. With this approach, a separate pthread is assigned to each data item to be extracted, each periodically executing the *Send_ISO15765Message* according to its configured data collection frequency. A further pthread is dedicated to receiving the diagnostic messages from the vehicle ECU using the *Receive_ISO15765Message* function.

The receiving pthread uses a shared memory to transfer the diagnostic data obtained from the ECU to its main process. Since this memory is implemented as a circular buffer of a limited size, it is crucial to ensure that the receiving phtread does not add data to the buffer when it is full, and that the main process does not remove it when the buffer is empty. This constitutes a classic producer-consumer problem, which is resolved using mutexes: mutexes synchronise the two processes, allowing them to access the buffer in mutual exclusion. In particular, we opted for a timed mutex, so that the add-on can notify Node.js when no messages are received from the diagnostic ECU, an event that is triggered when no data is made available in the shared memory within a certain time interval. In such cases, *ShareMeasurement.js* transmits only the GPS position to the MQTT broker and, if the event occurs repeatedly, suspends the sending pthreads under the assumption that the vehicle engine has been switched off. These pthreads are later reactivated following an interrupt on GPIO25. The source code for both pthreads and mutexes is contained in *Multithread.c*.

The native add-on based on *Extract_data.cpp* was developed using a different procedure from that employed for *Extract_VIN.c*, as this add-on must be executed asynchronously. Once launched, it performs data extraction in parallel with other dongle operations and communicates the collected data to *ShareMeasurement.js* as soon as it becomes available. To achieve this, we used the streaming-worker SDK, which allows the creation of a native add-on that runs in a separate thread and shares diagnostic data via a data stream, which Node.js retrieves through a callback. Since the streaming-worker module supports only C++, we had to include our C file in a wrapper, which explains the different file extension.

6.2.3.2 GNSS Positioning Software

Our device retrieves the GPS position of the ambulances with a dedicated add-on, obtained from the *GNSSModule.c* file via the N-API toolkit; this add-on first activates the GNSS engine of the Waveshare hardware and then acquires the spatial coordinates. Both these operations are performed through AT (Attention) commands – short strings of text that can be combined to perform tasks like dialling, sending SMS, or modifying connection parameters. The AT commands are sent to the Waveshare dongle over the Raspberry Pi's USB port, which also returns the result of the request to the main board. The outcome of this procedure is a string in NMEA (National Marine Electronics Association) format, which encodes the vehicle's location. This string also includes additional information, such as the number of satellites used to retrieve the GPS coordinates, the altitude of the antenna, and the timestamp of the measurement. For the algorithm for decoding latitude and longitude from the NMEA string, the reader is referred to external sources.

6.2.3.3 Software Cybersecurity

Before concluding this section, it is worth dwelling on the long-standing issue of dongle cybersecurity. From a cybersecurity point of view, the OBD port is indeed one of the most vulnerable attack points of the vehicle, as it provides direct access to virtually all its ECUs. It is then crucial to resort to a dongle that minimises threats to the vehicle, the driver, and the immediate surroundings. Regrettably, this is an extremely challenging matter, in light of the fact that the CAN protocol was not designed to be cybersecure, as mentioned in Section 6.1.2. Nonetheless, academic research proposes ingenious solutions to limit the vulnerabilities of the diagnostic port and shield the vehicle from the attacks that can originate from a compromised dongle. For more details on these solutions, which are beyond the scope of this book, the reader is referred to the studies available in the literature.

Although security is not among the core requirements for the dongle, our device still implements the cybersecurity strategies expected from a product of this kind. To begin with, the module for extracting diagnostic data is *secure by design*, allowing only ISO 15765 messages to be exchanged on CAN; it also limits the range of addresses from and to which these messages can be communicated, as well as the PID services that can be employed and the PIDs that can be extracted. The security of the JavaScript code is ensured, as far as possible, by the Obfuscator module, which makes the Node.js modules difficult to understand; the same problem does not arise for the add-ons, which are in the form of binary files. Higher-level security strategies can also be implemented. For example, the vehicle ID that the operators use to start a shift should never be presented in plain text on the dongle's screen but hidden behind a dynamically generated QR code; this ensures that only users with physical access to the dongle can start a shift via the app.

To further strengthen the cybersecurity of the dongle, it is finally recommended to use Ubuntu Core, a version of the Linux OS specifically designed for IoT and embedded applications. It offers, in fact, extremely valuable security strategies, from full-disc encryption to read-only root file systems, and employs digital signatures to guarantee the integrity and provenance of software. Furthermore, Ubuntu Core also allows transactional mission-critical OTA updates and even supports a real-time Linux kernel as a beta feature, which would provide deterministic response times to dongle operations.

6.3 CHAPTER SUMMARY

This chapter has demonstrated the centrality of system requirements in the choice of the technology to be employed in the implementation of a vehicle interface for fleet management applications. In fact, although it may seem reasonable to rely on an automotive data service provider to obtain vehicle data without resorting to an aftermarket device, this chapter illustrates how the non-functional requirements of the target applications lead instead to the use of a more conventional OBD dongle. It also proves the need for the ISO 15765 diagnostic protocol, which offers the highest level of dependability on an emergency fleet among the protocols in the OBD-II suite. Finally, although this component is not inherently innovative, the chapter demonstrates the importance of departing from existing dongles in both hardware and software design to ensure that the vehicle interface meets the demands of a context-aware, real-time fleet management application. Chapter 7 discusses the user interfaces, the last class of components to be described before proceeding with the testing of the system's functionalities.

7 User Interfaces

Any exchange of information between the CVP and the various entities interacting with it is enabled by interfaces, which can be seen as the system's boundary to the outside world. A CVP for ambulance fleet management requires multiple interfaces, since there are several classes of end users interacting with the system. Call handlers, for example, resort to a web application to manage fleet operations, while ambulance operators make use of a mobile app, which allows them to start a shift on a vehicle and receive notifications about changes in its operational status. The reasons behind the adoption of both the web and the mobile application can be found in the section "System Users", which provides an introductory overview of these components. Needless to say, there is no novelty in relying on user interfaces of this kind; in fact, it is difficult to think of a single fleet management application that does not feature either a web or a mobile application. Yet, the user interfaces that result from this project prove to be ground-breaking from a functional standpoint. The mobile app, for instance, allows the sharing of context data – never included in traditional CVPs – while the web application supports the automatic dispatching of ambulances, a utopia in the current landscape of emergency vehicle management.

The aim of this chapter is to analyse the design of the CVP user interfaces necessary for effective management of the NHS ambulance fleet: the mobile application is discussed in the section "Designing the Mobile Application", while the web app is covered in the next. It is of utmost importance to point out that the functionalities of these user interfaces can be achieved with different combinations of programming languages, architectures, and frameworks; as a result, this chapter, in contrast to the previous ones, largely focuses on the design of the components, providing only a brief outline of their underlying technologies.

7.1 DESIGNING THE MOBILE APPLICATION

The mobile application is a component of the CVP intended for ambulance operators, who use it to provide the data platform with information about their ambulance shifts, as well as to receive notifications of any changes in the operational status of their vehicle. The above unveils the dual nature of the mobile application, which operates both as a data consumer, since it provides users with information relevant to their activities, and as a data producer, in that it acquires context data that is not used immediately, but rather stored in the database and later retrieved to support decision-making. The ambulance operators, for example, are required to specify their clinical role before accessing the app's functionalities; this might lead the reader to assume, erroneously, that this user interface enables a different subset of features according to the type of operator that interacts with it. Instead, this information is simply retrieved so that it can be stored in the database and accessed at a later point in time; for instance, the role of the ambulance operator is used in the decision-making process to quantify the overall ability of the various ambulance crews that are currently on duty:

DOI: 10.1201/9781003688549-7

the resulting information enables the dispatching algorithm to automatically filter out vehicles that are not able to serve a particular emergency from a clinical perspective.

Since its use is restricted to accredited users only, the mobile application requires both a sign-in procedure, through which the ambulance operators can register on the data platform, and a login procedure, which allows them to access the app's functionalities by providing their credentials. The sign-in procedure allows the mobile application to acquire the personal information of the operator and forward it to the API, which is responsible for creating a new user profile in the database. In addition to their details, the operator must also provide their credentials, in the form of a staff number and a personal password; the former is the unique identifier that every staff member is already provided with, the latter a secret alphanumeric string chosen by the operator that the app employs for authentication purposes. In fact, the operator is required to submit their credentials to the mobile application when logging in, enabling the data platform to verify their identity by comparing the information provided with that stored in the database. If a valid match exists, the app grants the ambulance operator access to their personal page. Clearly, all data exchanges between the mobile application and the database, both the ones mentioned above and those that follow, are carried out through the API via HTTP, as already outlined in Chapter 5.

The personal page serves as the main hub for the functionalities of the mobile application, as it provides access to both the ambulance shift management functions and those for handling the calls that the vehicle is called upon to serve. Ambulance operators can *start an ambulance shift* from their personal page by specifying the ID of the vehicle they intend to operate on, which is displayed on the screen of the dongle that the chosen ambulance is equipped with. This interaction with the mobile app generates a new database record in the shift table, so that the system – unlike the existing CVPs – can account for this information during the various fleet management activities. In addition, the app subscribes to the MQTT topic of that ambulance, so that the user can receive automatic notifications about that vehicle while their shift is open. As a result, the app, downstream of the dispatching procedure, can promptly obtain information such as the destination to be reached, the urgency of the call, patient data, and any other details to support the emergency.

The personal page of the mobile app also provides the ambulance operators with the ability to *close the call that is being served* with the simple touch of a button. This interaction with the mobile application leads to the manipulation of the corresponding call record in the database, to which that API adds the type of service the emergency has been handled with – either see&treat or see&convey – and an end-of-call timestamp; the system is thus informed of the renewed availability of the vehicle. Along with this, the app publishes a message on the MQTT topic of that ambulance, so that all the staff members operating that vehicle can be notified of the completion of the intervention: this is an asynchronous communication that allows all the mobile apps with a subscription to the topic associated with the ambulance to instantly update their content, without having to periodically check the status of the call in the database.

Finally, the operators can use the mobile app to *end their ambulance shift*; this interaction results in the manipulation of the shift record that has been previously created, to which the API adds an end-of-shift timestamp. It also leads to the withdrawal of the subscription to the MQTT topic, since the user is no longer interested in the activities of that ambulance. The same outcome is obtained when the end user *logs out*, which can be considered as an explicit request to cease all vehicle activities.

A graphical summary of the functionalities of the mobile application can be found in the use case diagram of Figure 7.1, which contains some of the information that has already been included in the use case diagram of Chapter 3. This diagram pays special attention to the relations between the various use cases, since, as explained above, some of the functionalities of the app are enabled for the user only upon completion of other operations. For a graphical representation of the tasks that are behind the single app functionalities, the reader is referred to the sequence diagrams that follow. These are high-level diagrams describing the way the mobile app, in the form of a black-box component, interacts with the user and with the rest of the CVP; information on the operations performed by the various software modules that the app is composed of, which have not yet been introduced in this book, is therefore omitted. Some of these details, however, can be extrapolated directly from the code included in Appendix D. For information on the manipulations that the API performs on the database data as a result of the interactions of the ambulance operator with the mobile app, the reader is instead referred to the code listings of Appendix B (Figures 7.2–7.6).

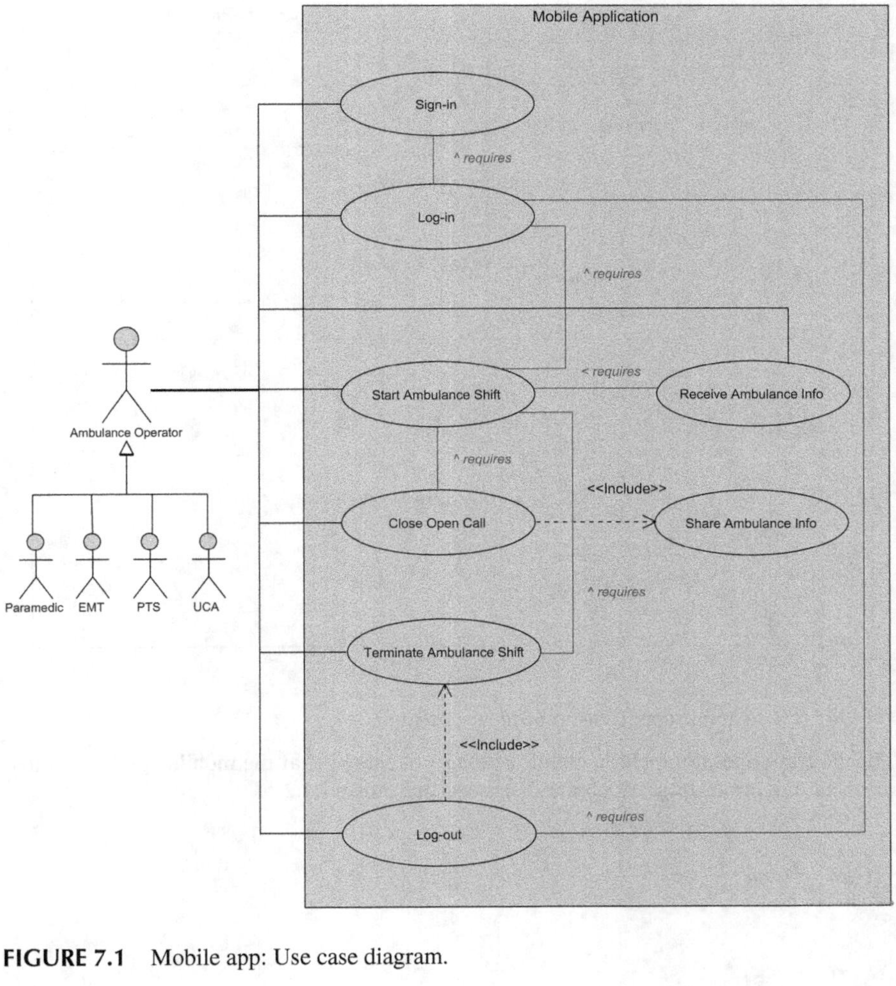

FIGURE 7.1 Mobile app: Use case diagram.

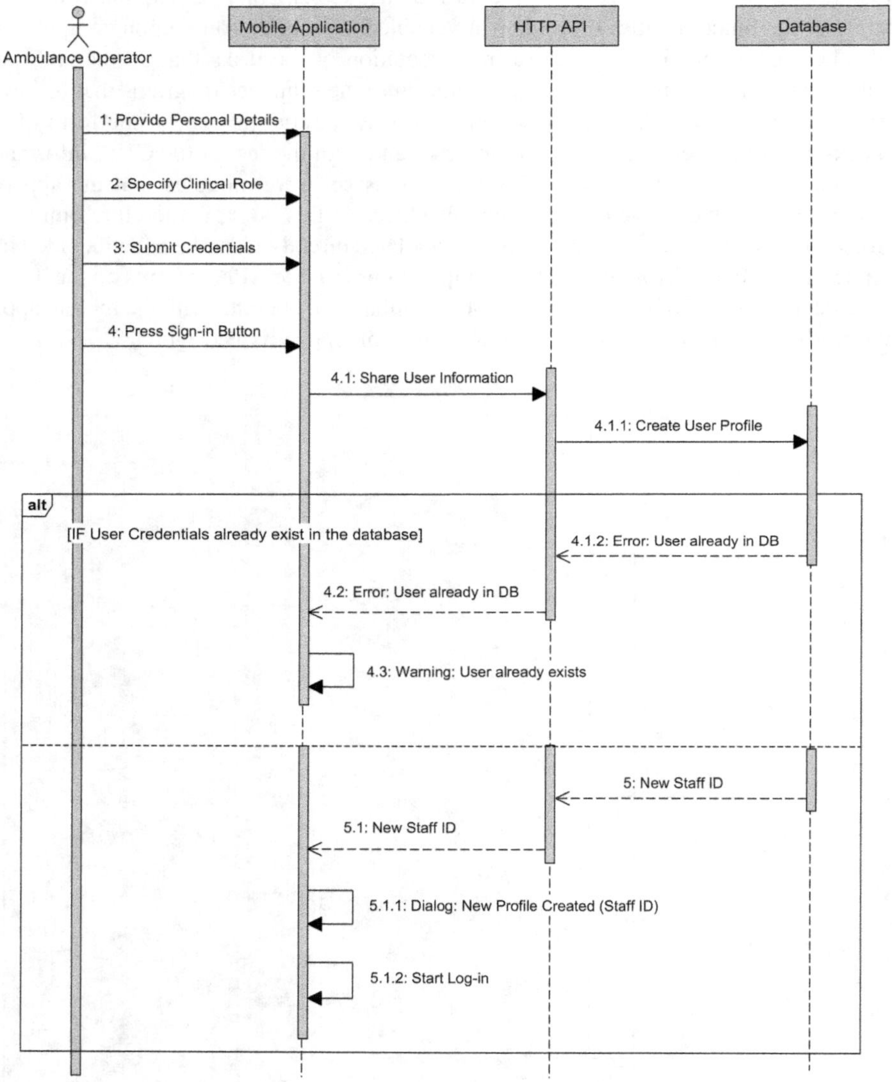

FIGURE 7.2 Mobile app, sign-in: Sequence diagram.

The diagram does not include details of the local checks that the mobile application performs on the input data, which are discussed in Section 7.1.2.

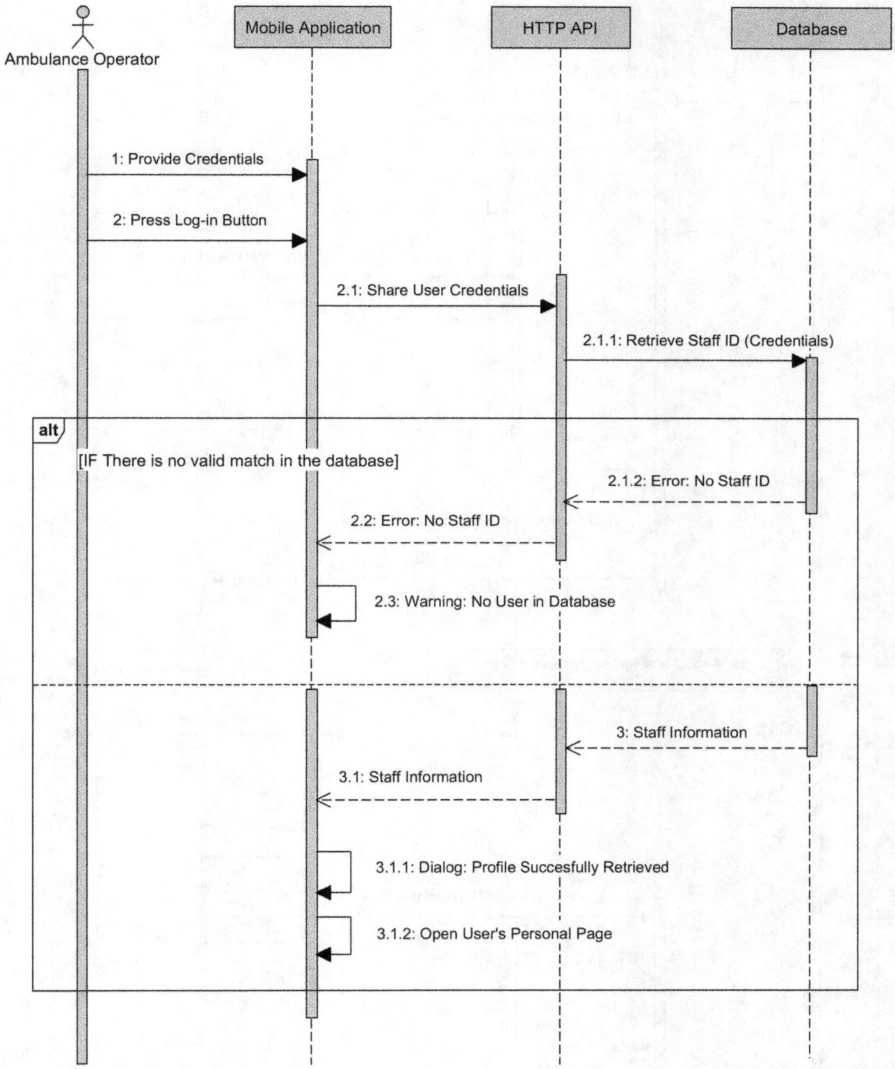

FIGURE 7.3 Mobile app, login: Sequence diagram.

The diagram does not include the operations for storing the user data in the device's memory as shared preferences, which are discussed in Section 7.1.2.

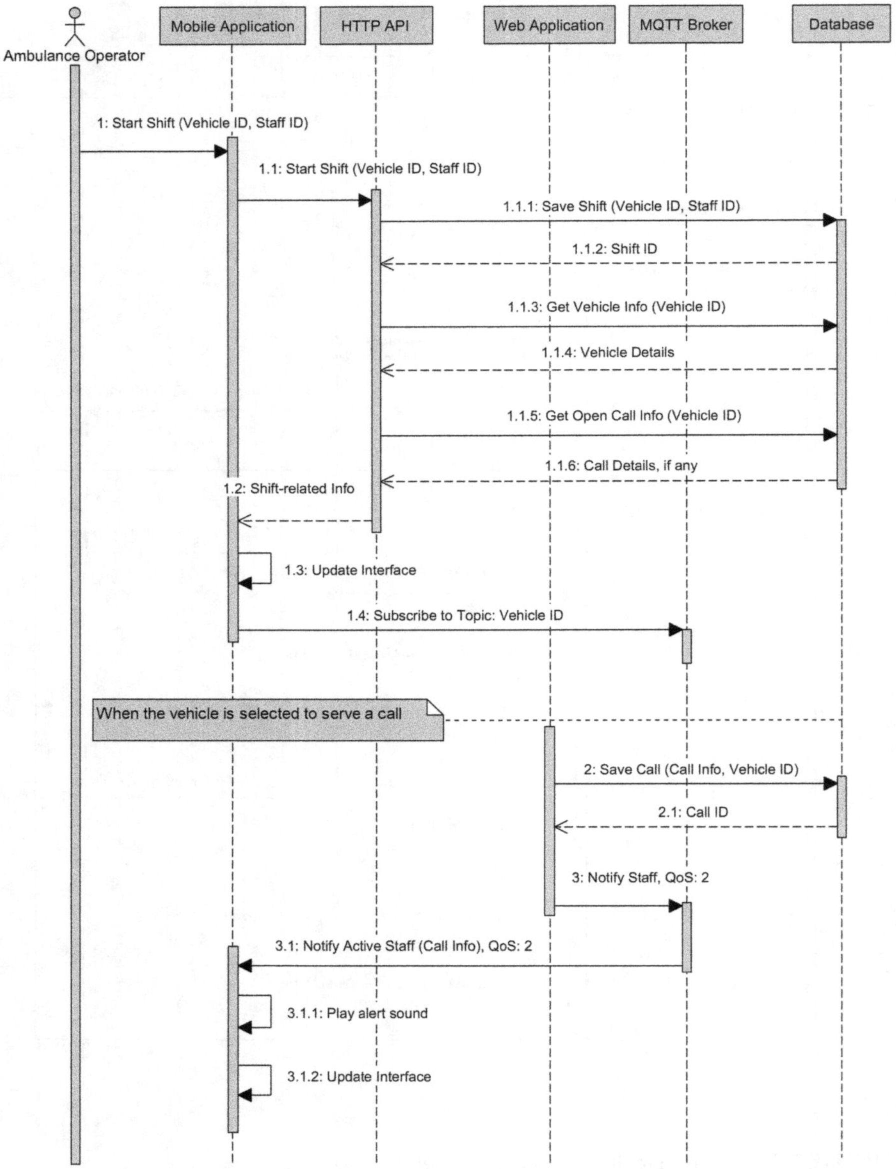

FIGURE 7.4 Mobile app, open shift: Sequence diagram.

The diagram splits the queries for obtaining vehicle and call details for mere legibility reasons; this information is, in fact, obtained using a single query via an SQL JOIN.

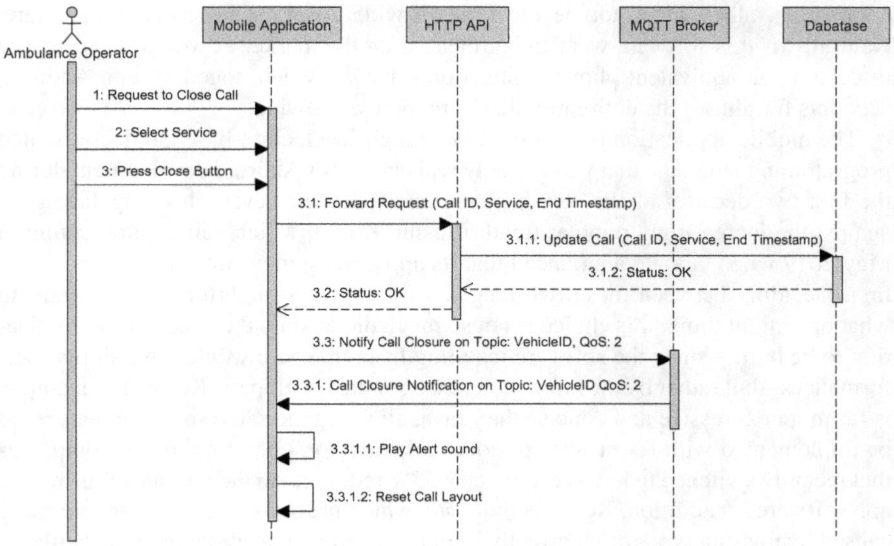

FIGURE 7.5 Mobile app, close open call: Sequence diagram.

It should be noted that the broker notifies all mobile apps with an open shift on the vehicle responsible for that emergency, including the one that sent the MQTT message.

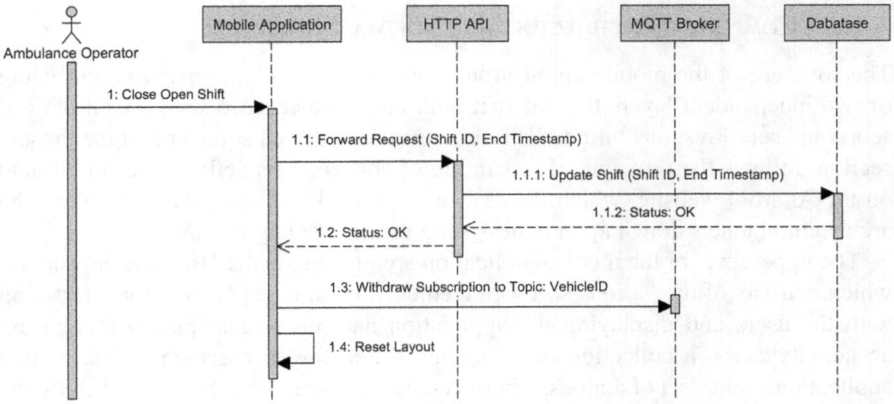

FIGURE 7.6 Mobile app, close open shift: Sequence diagram.

The same interactions occur when the operator logs out, followed by the removal of the user's credentials from the shared preferences and the exit from their personal page.

7.1.1 Underlying Technology

The mobile application is intended for *Android* devices only. The reasons behind our decision to tie the software to a single mobile operating system stem from the observation that Android is by far the most popular mobile OS on the planet, holding a share of 71.90% of the global market; this design choice ensures that the mobile application, albeit in prototype form, has a wide coverage in terms of hardware. Naturally, if this software were to be released on the market, it would be necessary to develop an equivalent phone application for iOS, which, together with Android, accounts for almost the entire mobile OS market worldwide.

The mobile application is built on *Java*, a high-level, class-based, object-oriented programming language that was the only real choice for Android development during the first two decades of the new millennium. Today, however, choosing Java goes against the increasingly popular trend of using Kotlin, which gained momentum in May 2019 when Google announced that its app development would become Kotlin-first. Deciding between these two languages is not so straightforward; contrary to what one might think, this choice is not so much dictated by the needs of the application to be built – since the software resulting from their use would have similar performances – but rather by the mere preferences of the developers. Kotlin, for example, is far more expressive and concise than Java, allowing the same software features to be implemented with fewer lines of code. TripAdvisor, one of the many companies that recently switched to it, has estimated a 33% reduction in the volume of its mobile app software. In addition, Kotlin is *null safe*, which prevents errors that are normally caused by missing types or incorrectly initialised parameters, reducing the likelihood of application crashes by 20%. Yet, despite these advantages, our preference for a more mature technology, the need for greater flexibility – especially in terms of third-party libraries – and the fact that the back-end of our web application is also built around Java, drove us towards a more conservative choice.

7.1.2 Component Architecture and Software Structure

The software of the mobile application adopts a *two-level architecture*, consisting of two independent layers that interact with one another, each with a well-defined scope and set of responsibilities. This architecture, outlined at the end of the present section, follows the principle of separation of concerns; its definition can be found on the Android website, which the reader is referred to for an overview of all the architectural tenets drawn upon in designing the mobile application.

The upper level of the mobile application architecture is the *User Interface layer*, which consists of the various software elements that are responsible for interacting with the users and displaying the application data on screen; placed therein are an activity class, a collection of fragments – one for each screen of the mobile application – and a set of dialogs, whose details are discussed in Section 7.1.3. Below this is the *Data layer*, which contains the application data together with its business logic, that is the algorithms that determine the way this data is to be created, stored, and processed. This layer includes a group of model classes, one for each of the entities involved in the application – namely Staff, Staff Role, Vehicle, Shift, and Call – which are modelled on the corresponding database tables. The communication between the elements of these two layers, which takes place in both directions, is achieved via the *observer pattern*: this design choice ensures that the fragments – the so-called observers – can register with and receive automatic notifications from the observables, which are the model classes, without the latter having to maintain any information about them. This ensures a sharp separation between the two layers of the architecture and avoids circular dependencies between their elements.

As can be observed from the component diagram in Figure 7.7, the data layer also includes a variety of *utility classes*, primarily intended to enhance the user experience of the mobile application. Among these is the *SharedPreferencesUtil*, a class that allows the software of the mobile app to manage the local persistence of the user's credentials; this information is stored in the device's memory following the login procedure and retrieved by the mobile app each time the user wishes to access their personal page; the ambulance operators are thus relieved of the need to provide their staff number and password every time the application is to be started. This way, the authentication is only required after the log-out, which removes the user's credentials from the shared preferences file.

Another utility class of paramount importance is *CallMonitoring*, which defines a foreground service through which the smartphone can receive MQTT messages even when the app is not active on the screen. This class is a significant contributor to the reliability of the app, as it allows the ambulance operators to receive vehicle status notifications even when the screen of their device is locked or while they are using other phone functionalities; those just mentioned are, indeed, quite plausible scenarios, since the device the app is installed on is the personal phone of the ambulance operator.

Among the utility classes is also *SupportFunctions*, which contains a number of functions shared by the various software modules of the mobile app. Noteworthy is *DelayedNavigation*, a function that, as its name plainly suggests, slightly delays fragment navigation, making the user interactions with the app more pleasant.

The interaction of the mobile application with the rest of the data platform is entrusted to the data layer, regardless of whether this user interface takes on the role of data producer or that of data consumer. The software of the mobile app is equipped with a further group of classes, aimed at simplifying the communication between the data layer and the outside world. One of these is *VolleyDispatcher*, which allows the model classes to communicate with the database via HTTP; it is based on the Volley library and modelled after the singleton design pattern, so as to instruct the software of the app to create exactly one instance of RequestQueue throughout the entire application life cycle, which optimises the usage of the phone's resources. By contrast, the communication between the foreground service and the broker – both in transmission and reception – is delegated to *MQTThelper*, a class built on the Eclipse Paho library; the data that the foreground service obtains from the MQTT broker is then forwarded to the main activity of the app via a broadcast receiver in a publish-subscribe – like fashion.

As is evident from the above, the mobile app primarily operates as a data bridge to the rest of the CVP, ending up resorting to a software with a rather simple business logic. Our decision to opt for a two-level rather than a three-level architecture stems precisely from this, which resulted in the choice of relegating algorithms to the data layer instead of placing them in a dedicated intermediate software level. Should the decision be made to move from a prototype to a commercial product, the recommendation is to take this implementation as a mere starting point and to restructure the software using an architectural pattern like MVC, MVP, or MVVM, for whose details the reader is referred to external sources; such a design would provide the software with superior readability, maintainability, and testability properties, factors that should certainly not be underestimated when implementing a mobile application of this significance.

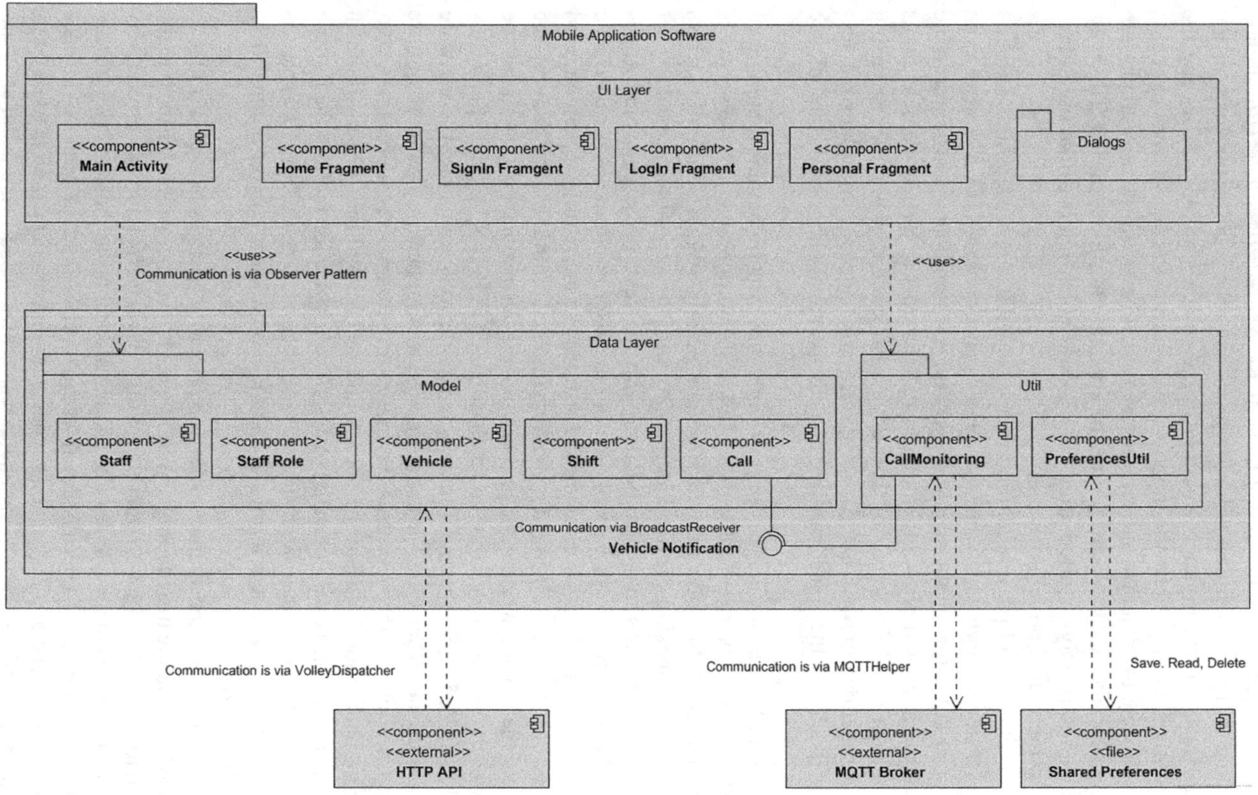

FIGURE 7.7 Mobile app, software architecture.

This component diagram provides a non-exhaustive view of the software structure, as it omits a number of its components and some of the interfaces, relationships, and dependencies between its modules.

7.1.3 UI Design and Component Functionalities

The user interface (UI) layer of the mobile app consists of a main activity, in charge of the window where the app screens are drawn, and a set of four fragments, each dealing with the layout of a specific screen. This is a design inspired by the *single-activity pattern*, which aims to relieve the activities of all the business logic related to the UI, delegating these components to the mere hosting of fragments: the activity is thus limited to displaying the correct screen, while the designated fragment manages its layout. Adopting this paradigm provides several benefits, including the possibility of taking full advantage of the Navigation Jetpack library; this library simplifies tasks such as managing the transitions between app screens and their animations, which can be a major challenge when building the UI layer with activities alone. This paradigm also guarantees a notable performance gain, as the life cycle of the fragments does not require the intervention of the OS, unlike that of the activities. Even sharing data between app screens becomes considerably easier than when resorting to activities only; this is due to the ViewModel component, which, being linked to the life cycle of the activity, is able to survive fragment replacement. Clearly, no such component is required for our app, since the only data shared between fragments is the operator's credentials, which must be accessed via the shared preference file.

A visual representation of the logic for navigating the screens of our mobile app – obtained by using the Navigation library – is presented below in Figure 7.8; as can be seen from the navigation graph, the app has been named VEICOLNET, in line with what has already been done with the OBD-II dongle. The layout of the individual screens, along with the operations that the software performs when the user interacts with them, is briefly discussed in the pages that follow (Figure 7.9).

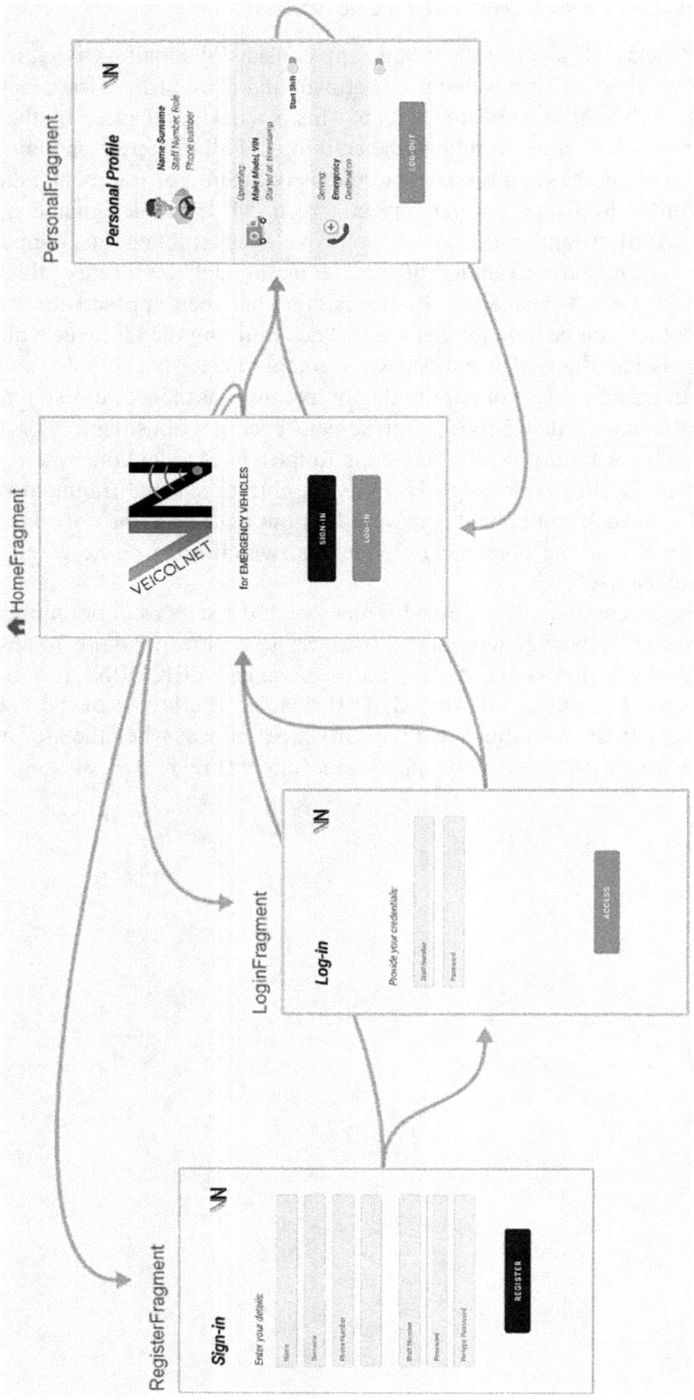

FIGURE 7.8 Mobile app, screen navigation.

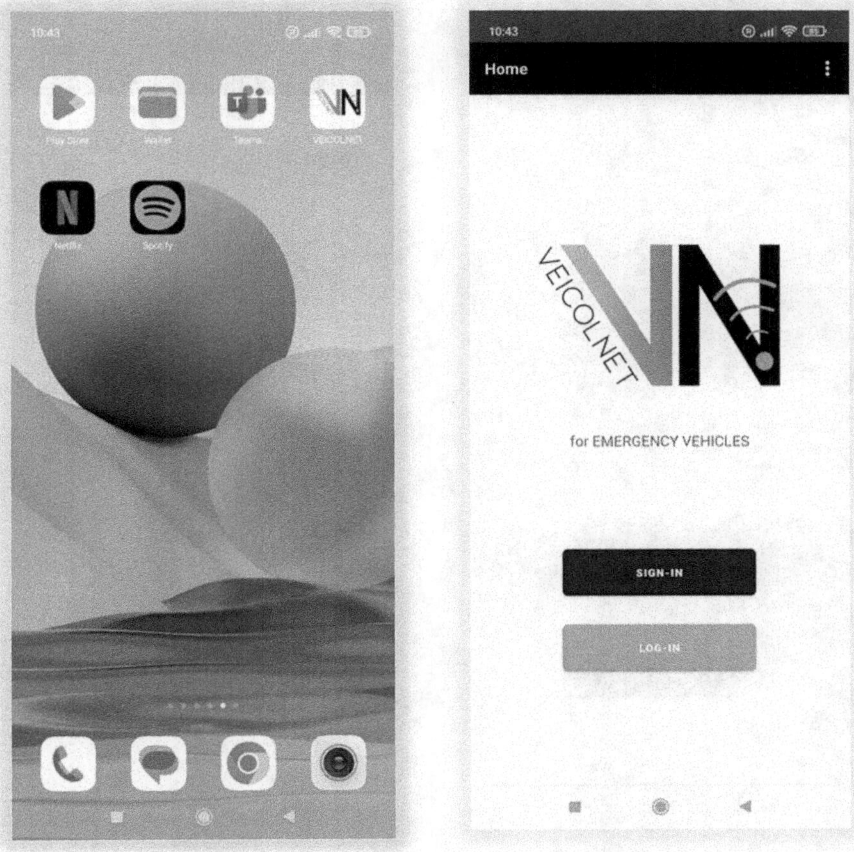

(a) (b)

FIGURE 7.9 Mobile app, layout: Icon and home screen. (a) App icon and (b) home screen.

Home Screen is the screen the ambulance operator is directed to when clicking on the corresponding *icon* in the app gallery of their smartphone. As can be seen from the navigation graph, the home screen of the app is connected to all the other destinations of the UI: if the operator does not possess a personal profile yet, they can navigate to the sign-in screen, alternatively to the login screen, both of which can be retrieved by clicking on the *dedicated button* of the layout.

Although less obvious, direct navigation between the home screen and the personal page of the ambulance operator is also possible. This transition occurs seamlessly when the user reopens the app without having logged out, after the fragment retrieves their credentials from the shared preferences file. Naturally, each screen of the mobile app also includes an outgoing transition back to the home screen (Figure 7.10).

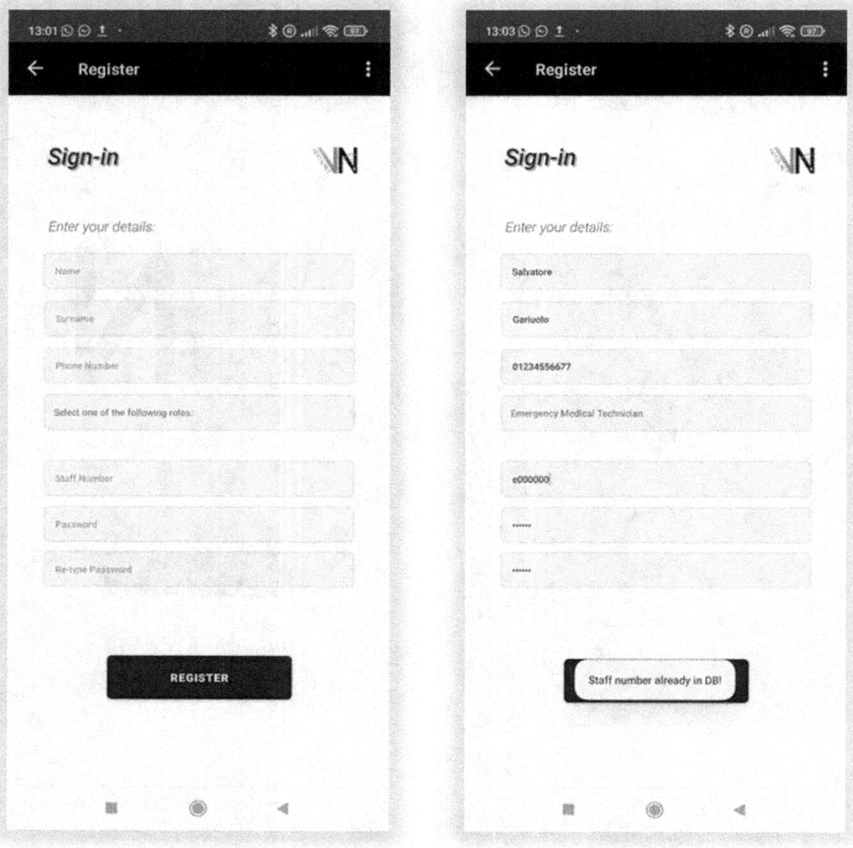

(a) (b)

FIGURE 7.10 Mobile app, layout: Sign-in screen. (a) Empty layout. (b) Warning message.

Sign-in Screen is the screen of the UI that is responsible for creating the operator's profile. It features a linear layout with a form-like design, consisting of six *text edits* and a *spinner*, which lists all the roles of ambulance operators. The spinner is automatically populated through an HTTP request to the API when the view is created, so as to avoid having to hard-code this information within the software. Clicking on the *register button* triggers a further HTTP request, which is, however, only forwarded if the operator has completed all the fields of the form and if the two passwords provided match. The data contained in the request message is used by the API to create a new record in the staff table, whose outcome is returned to the mobile app: if this database manipulation fails – in case the user's credentials already exist in the data storage – the sign-in fragment displays a *warning message* via a toast on screen; otherwise it directs the ambulance operator to the login page (Figure 7.11).

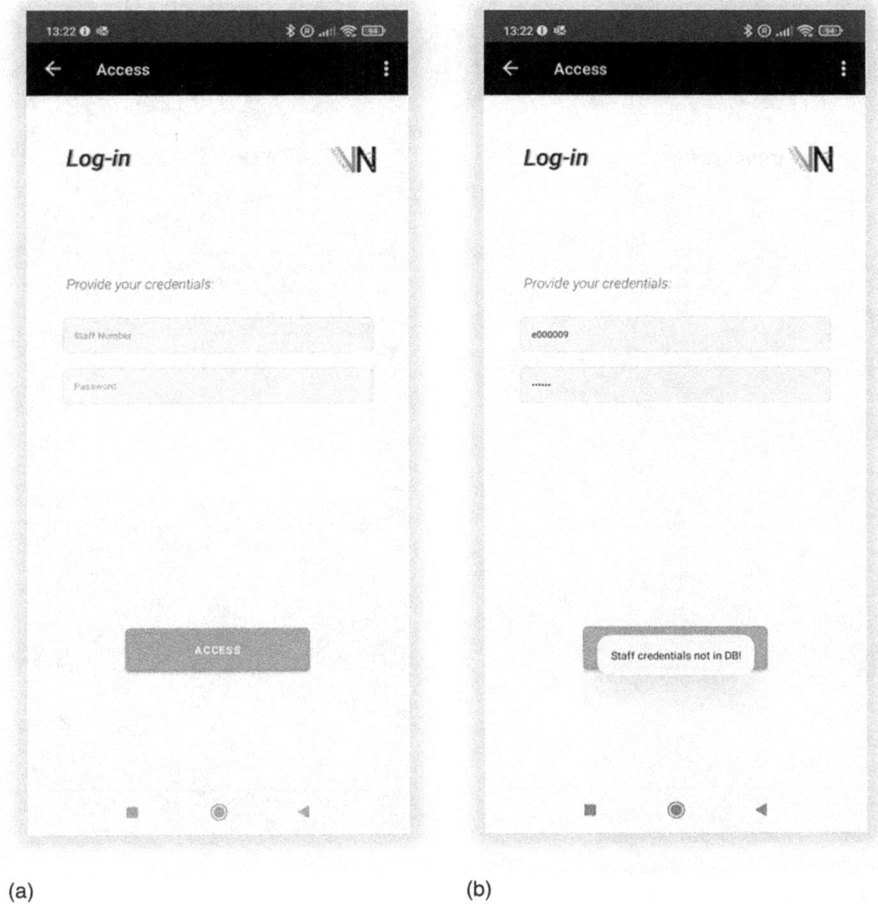

FIGURE 7.11 Mobile app, layout: Login screen. (a) Empty layout. (b) Warning message.

Login Screen is the screen of the UI that allows the ambulance operator to access the functionalities of the app by providing their credentials. It relies on a rather trivial form, which invites the operator to enter their staff number and personal password. Clicking on the *access button* initiates the user authentication procedure, which takes the form of an HTTP request to the API, where the latter compares the data provided with that stored in the database. Should the credentials of the user match, the API returns all the ambulance operator's details, which the login fragment stores in the local *shared preferences* file in the form of key/value pairs. This is automatically followed by the transition to the personal page of the operator. If the credentials provided are not characteristic of any database record, the data returned by the API prompts the fragment to display a *warning message* on screen, urging the user to review the credentials that have been entered (Figure 7.12).

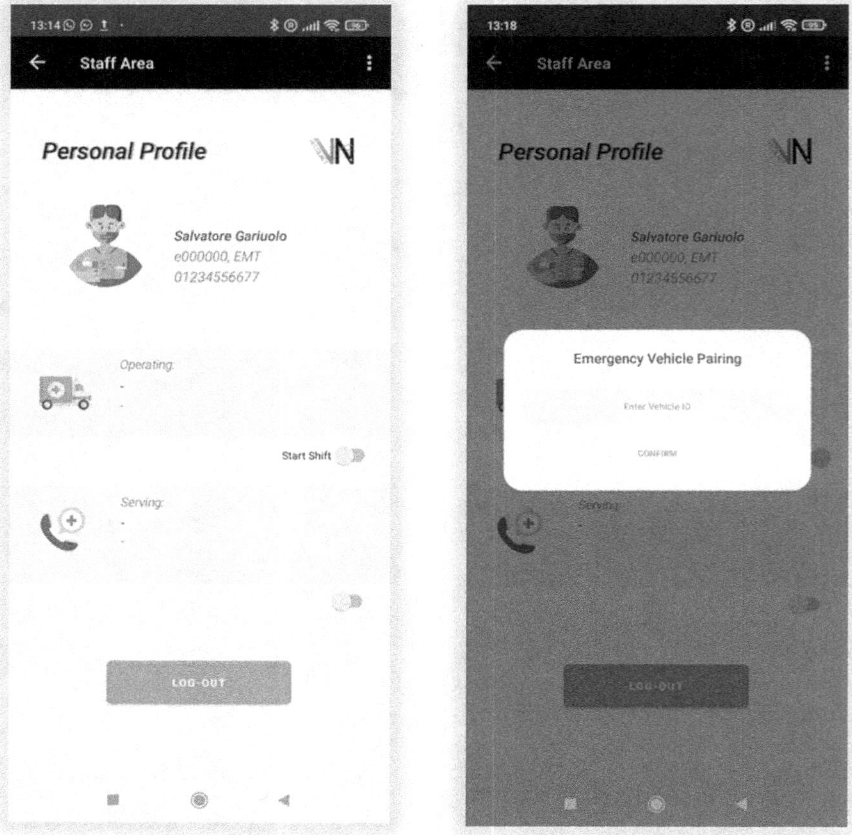

(a) (b)

FIGURE 7.12 Mobile app, layout: Personal page. (a) Empty layout. (b) Start shift dialog.

Personal Page is the screen of the UI that provides access to the main functionalities of the mobile application. It consists of three different sections. The first, located at the top of the screen, displays the user's details alongside an *icon* chosen according to the operator's role; this information is automatically retrieved from the shared preferences file when the view is created. Immediately below are the two other sections: one for managing ambulance shifts, and the other for managing emergency calls. After logging in, the user is only allowed to interact with the former through a *switch*, which triggers the shift-start procedure: interaction with this widget opens a *dialog window* where the operator can provide the ID of the vehicle to be associated with. This results in an HTTP request to the API, which not only creates a new shift record in the database but also retrieves information on the ambulance and its call, if any (Figure 7.13).

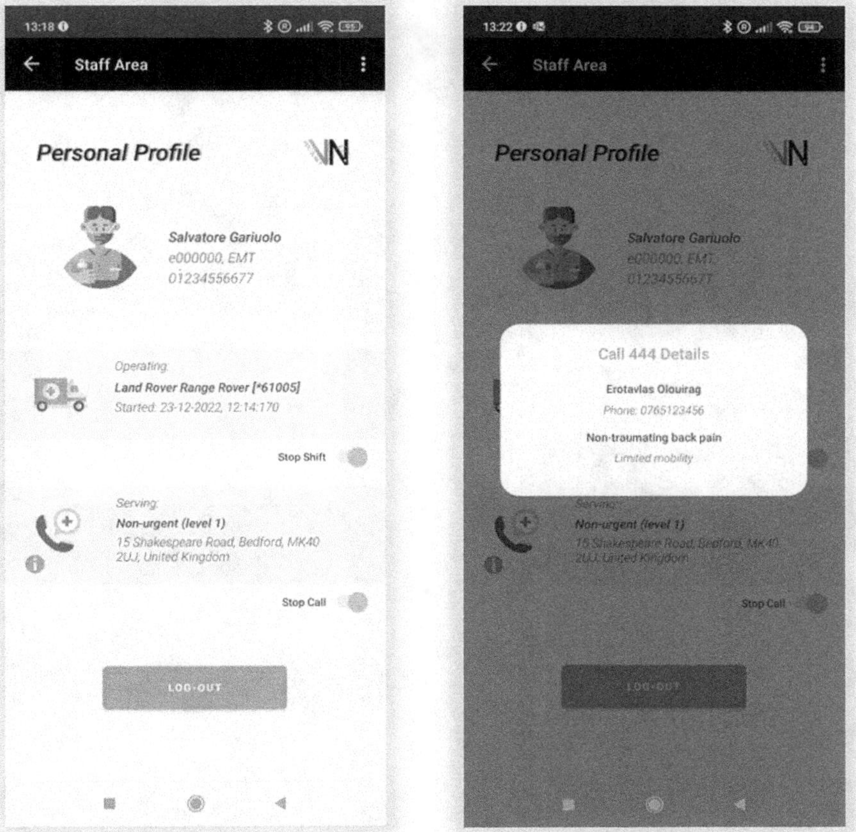

(a) (b)

FIGURE 7.13 Mobile app, layout: Call info. (a) Call info and (b) additional details.

Once received, vehicle details are displayed in the relevant section of the layout, while call-related information is split between the personal page and a dedicated dialog, which can be accessed by clicking on the *info icon* that appears on screen. Call information can be received either via HTTP, if the ambulance already has an active call when the vehicle shift is started, or over MQTT, with the receipt of a dispatch notification. In fact, while a vehicle shift is in progress, the mobile app relies on a foreground service to receive notifications about any changes in the operational status of that ambulance. As a result, the personal fragment can automatically update the section of the layout related to the call when a new MQTT message is received. Some of this information can also be displayed in the *status bar* of the foreground service, as shown in Figure 7.14a; this way, the operator can access key call details without having to unlock the smartphone screen.

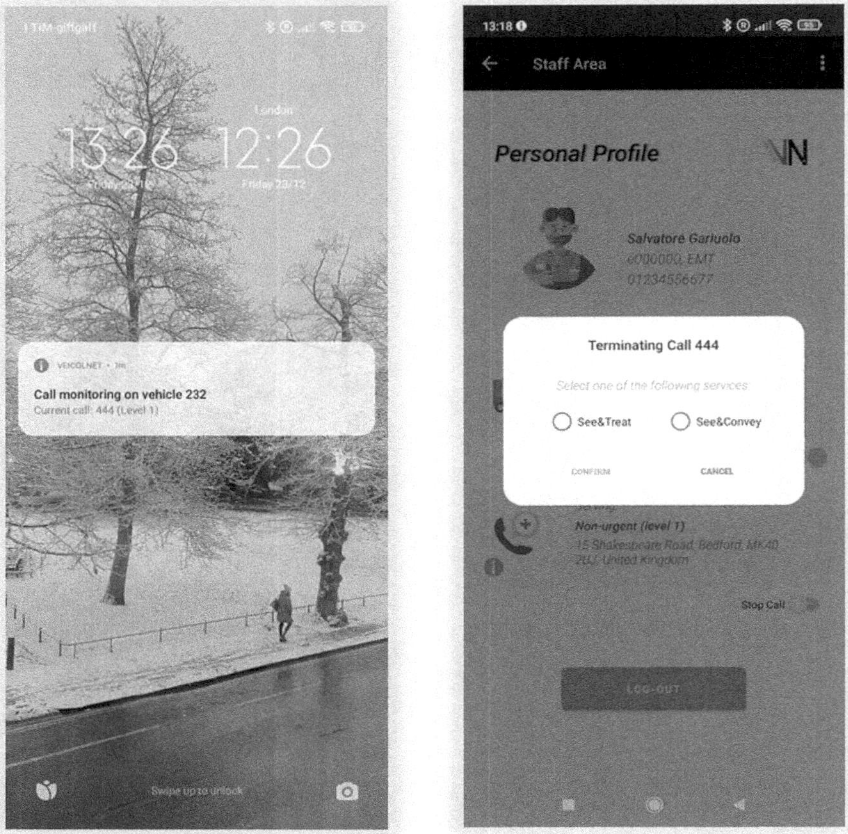

(a) (b)

FIGURE 7.14 Mobile app, layout: Vehicle monitoring and call termination. (a) Foreground service. (b) Close call dialog.

If there is an active call, the app allows it to be terminated via a dedicated *switch*; when interacting with this widget, the fragment displays the *dialog window* shown in Figure 7.16b, which allows the ambulance operator to determine the type of service that the emergency has been handled with. The personal fragment can then fire an appropriate request to the API and send a call closure notification via MQTT; yet, the call-related section of the layout can only be cleared when the app receives this message back from the broker. An open shift can be closed, instead, with the same *switch* that allowed the user to start it. Interacting with this widget, while a shift is open, results in an API request and in the withdrawal of the subscription to the MQTT topic of that ambulance; the fragment can then reset both the vehicle and the call sections of its layout. The same procedure is performed when the user clicks the *log-out button*, which brings the mobile app back to its home screen.

7.2 DESIGNING THE WEB APPLICATION

A CVP for emergency fleet management should feature two web applications, as it targets as many classes of end users, namely fleet managers and call handlers; our prototype, on the other hand, only includes the one intended for the latter. Our decision not to implement a web app for fleet managers, however, is not a superficial omission, but a deliberate choice dictated by the objectives of our research. The aim of this project, in fact, is not to develop a CVP for commercial use but to demonstrate that by tailoring the data platform to the non-functional requirements of the applications to be served, superior fleet management can be achieved. Yet, proving that this approach improves the operational efficiency of the fleet manager would have required the development of new, more sophisticated models for evaluating fleet performance, which is beyond the scope of this research; indeed, it is not the web app per se that influences the capabilities of the fleet manager, but the ability to leverage such models, which our CVP enables through context data.

The web app included in our system allows call handlers to *live monitor the status of the ambulance fleet* and to *initiate the dispatching procedure*, which selects the ambulance to be activated upon receipt of an emergency call. Needless to say, it is in the functionalities of this UI that all the advantages of using a tailor-made CVP become apparent. Fleet monitoring, for example, is no longer limited to vehicle tracking, but provides a detailed narrative of the status of the ambulances, displaying information on the crew operating them and the call they are active on. It is the dispatching, however, that undergoes the greatest upheaval; our CVP, unlike the existing systems, does not merely support the activities of the call dispatchers, but supersedes them: the ability to promptly access the full picture of information enables, in fact, an automated, quasi-immediate, and error-free decision-making.

Our web application allows the call handler to *assess the operational status* of the ambulance fleet by means of an extremely intuitive real-time monitoring function. The operational status of the fleet is, in fact, displayed on a map, on which the web app places an icon for each of the ambulances located in the geographical area of interest. This enables call handlers to quickly grasp the conditions of the fleet in a certain region, without having to browse through the details of the individual ambulances: each marker is, in fact, associated with a colour that is indicative of the operational status of the corresponding vehicle. The call handler can then access the detailed report of a particular ambulance by clicking on the corresponding map icon: this interaction prompts the web application to switch from multi-vehicle to single-vehicle monitoring mode, providing the end user with live information on the ambulance operators that are active on the chosen vehicle, as well as details about the call the ambulance is currently serving, if any.

Although primarily intended for fleet managers, our web application also includes a function for *live monitoring the internal status of the vehicles*. Similar to the function described above, this real-time monitoring function features both a multi-vehicle mode, whose marker colours reflect the emissions produced by the vehicles, and a single-vehicle mode, which displays the structural specs of the chosen ambulance, a real-time overview of its engine status, and a live estimate of its emissions.

The *ambulance dispatch* function included in our web app follows the same procedure used in the control centres of the NHS, with the call handler that interacts with the system to provide information on the emergency that has arisen, which is saved in the storage unit; this is followed by the selection of the vehicle to be activated, chosen according to the data made available by the CVP. However, unlike the existing commercial platforms, where this selection is carried out manually by the call dispatcher based on the geographic location of the fleet alone, our system relies on a fully automated procedure, which is clearly more responsive and less prone to errors. Moreover, the possibility of resorting to context data, as well as information from external sources, such as traffic data, enables a significantly more sophisticated dispatching procedure: our web app can select the available vehicle with the shortest time to reach the destination among those operated by a crew capable of handling that emergency. In other words, the ambulance dispatch is performed through an improved version of the *shortest route search algorithm* that is traditionally used for managing emergency fleets. Our decision to resort to this algorithm, although the platform allows for the innovative methods available in the literature, stems solely from the need to compare the performances of our CVP with that of commercial systems, which obviously requires a common ground in terms of procedures; naturally, this does not preclude, in the commercial version of the platform, the use of more efficient methods than the one adopted. Once the ambulance has been dispatched, the web app stores the outcome of the procedure in the database and notifies the staff operating that vehicle over MQTT. Finally, it provides the user with an estimate of the time needed for the ambulance to reach its destination, so that this information can be shared with the caller.

A graphical recap of the functionalities of the web application can be found in the use case diagram of Figure 7.15; for each of these use cases, a sequence diagram is also provided, illustrating the interactions that occur between the web application, seen as a black-box component, and the rest of the system. For details on the tasks performed by the different components the web application consists of, the reader is referred to the listings in Appendix D; this appendix also includes information on the corresponding database manipulations, since the web application, unlike the mobile app, does not rely on intermediate components like the API to interact with the data storage unit (Figures 7.16–7.18).

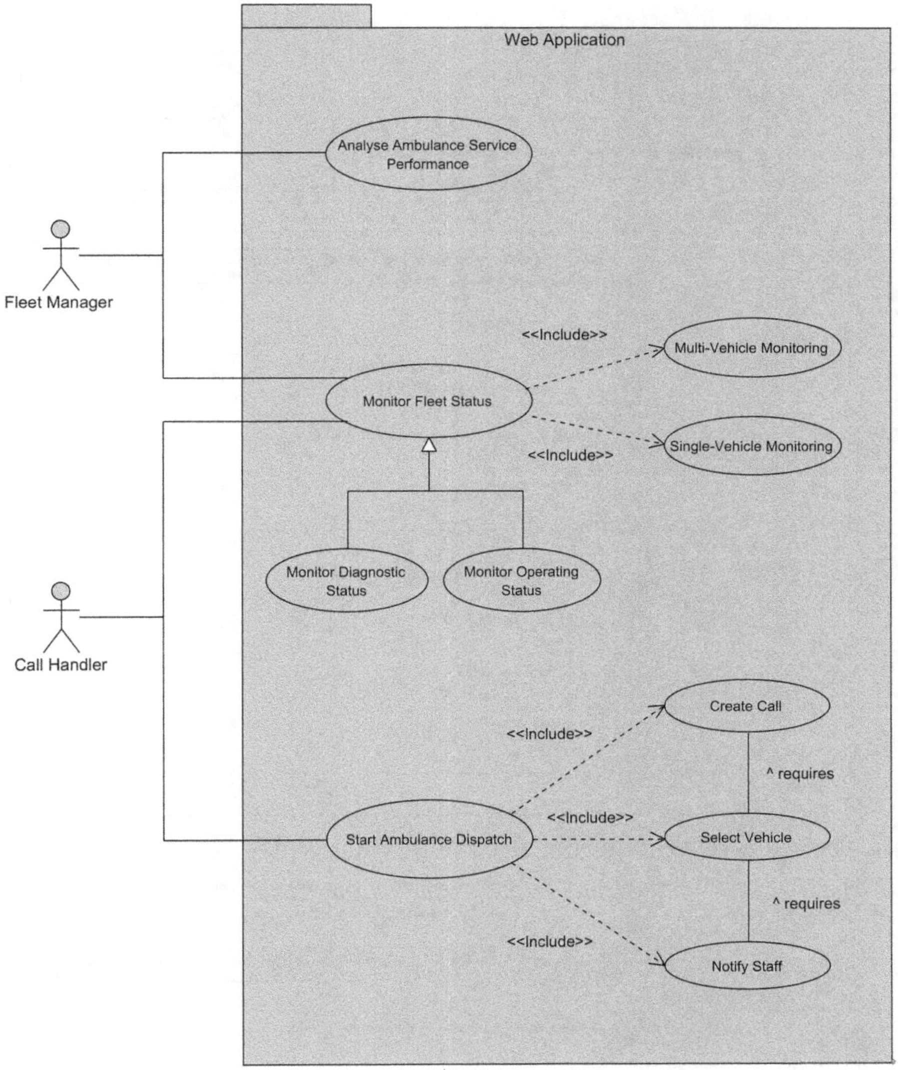

FIGURE 7.15 Web app: Use case diagram.

This use case diagram covers the functionalities of both web applications. Hence, it also includes the use case on vehicle performance evaluation intended for fleet managers, which is not implemented in the prototype version of the data platform.

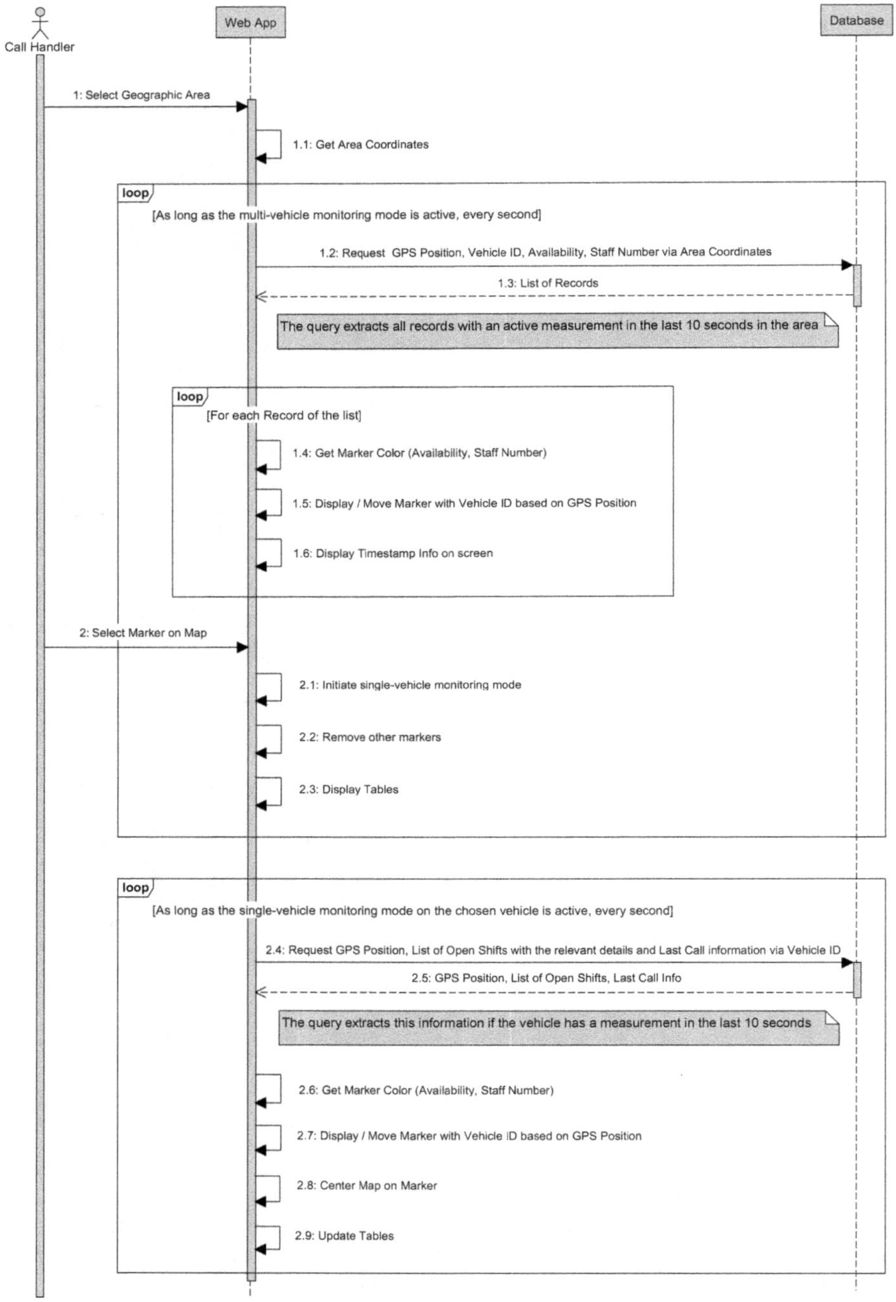

FIGURE 7.16 Web app: Fleet operation monitoring.

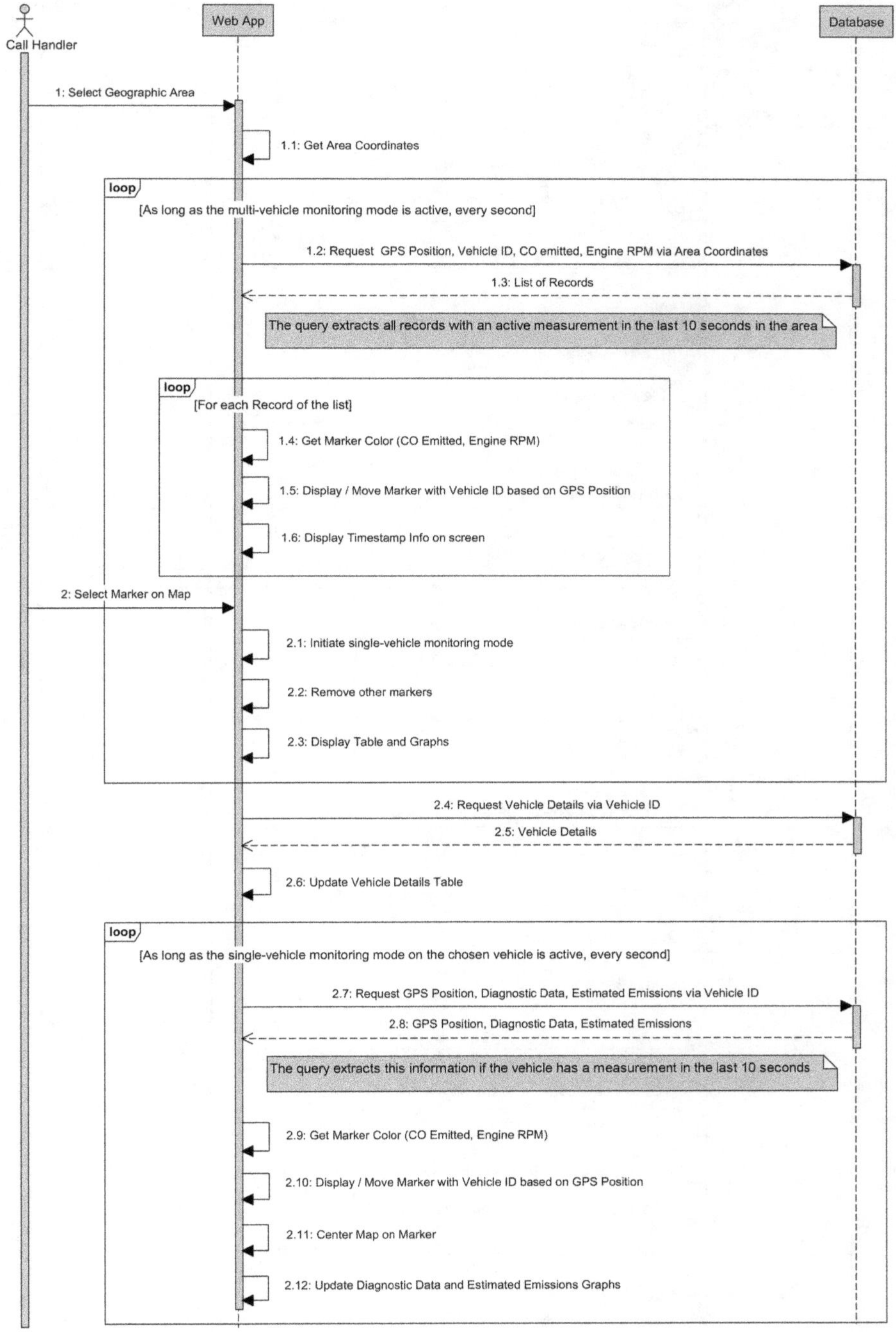

FIGURE 7.17 Web app: Fleet diagnostic monitoring.

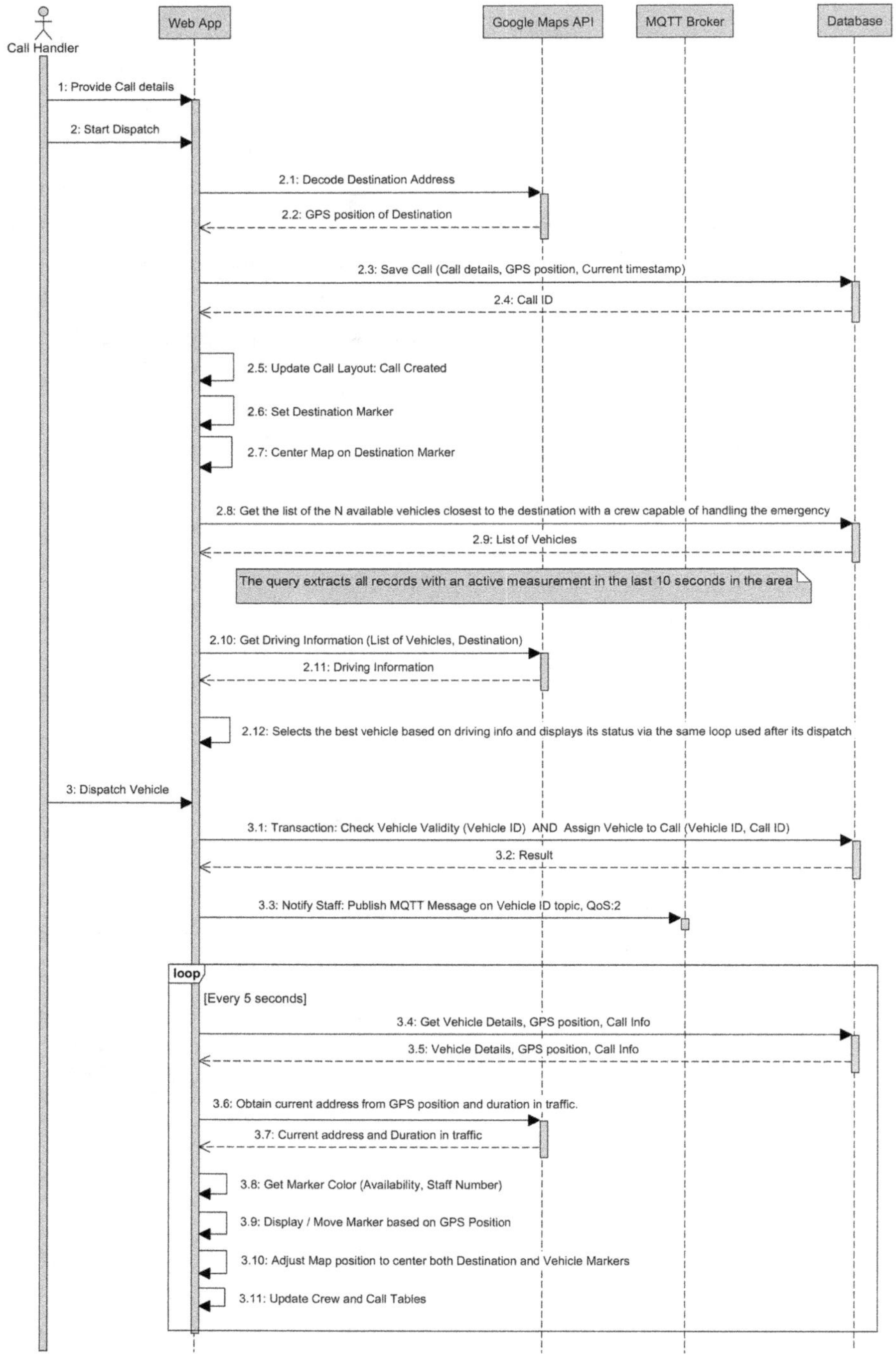

FIGURE 7.18 Web app: Supervised ambulance dispatch.

7.2.1 Underlying Technology

The web app has been built with *Ajax*, a set of web development techniques whose acronym stands for Asynchronous JavaScript and XML. As its name plainly suggests, Ajax enables the development of asynchronous web apps, in which the communication between the browser and web server takes place without interfering with the display of the web page content. This is achieved by decoupling clients and server, delegating the management of the communication between front-end and back-end, which takes place over *HTTP*, to the Ajax engine. The resulting advantage is the extreme responsiveness of the web app, which becomes capable of updating the information on screen without having to refresh the entire web page. Ajax is therefore ideal for applications whose content varies regularly, since the browser can request from the web server only the database records that change over time; this can guarantee a significant reduction in network traffic, as less data is exchanged between the browser and the web server. Even more remarkable is the impact on the user experience, which becomes smoother and less tedious, without the frustration of having to sit through the page being refreshed every few seconds.

Ajax has proved to be the obvious choice to support the fleet monitoring functions of our web app; both functions, in fact, involve frequent data requests to the web server, since they are designed to provide the call handlers with up-to-date information on the status of the ambulance fleet. Thanks to Ajax, our web application can only request the position of the vehicles in the geographical area of interest, which is used to update the position of the markers placed on the map, as well as the information to be displayed in the tables and graphs of the dashboard; the web app can thus avoid refreshing the entire web page every second, which would have made the entire data platform unusable. Although less obvious, Ajax is also used to support the ambulance dispatch; this topic, however, is not covered here, but in Section 7.2.3, since the considerations relating to this matter are easier to comprehend if accompanied by the description of the design of the web page.

As the name of the paradigm suggests, the technology that underpins the client-side of our web application is *JavaScript*, whose scripts are responsible for adding behaviour to the web page. These scripts rely on a number of JavaScript libraries, such as *jQuery*, which provides native methods for Ajax functionalities; this library proves to be extremely helpful since it also deals with event handling, *CSS* animations, and the manipulation of the DOM, which is the logical representation of all the elements of the web page in a tree-like structure. Unfortunately, all these scripts, being based on JavaScript, are immediately accessible with a simple inspection of the page, which undermines the privacy and security of the system. Therefore, as already suggested for the Node.js modules of the dongle, the code should be made unreadable through obfuscation techniques. The use of such techniques, however, is not sufficient to protect confidential data such as API keys, which still remain accessible from the browser; as such, any interaction with external APIs is done through the back-end. With regard to compatibility, it should be noted that all modern browsers have built-in support for Ajax, which ensures that our system can be accessed from any web-enabled device.

The language that the web server is based on is *Java*, while the format it employs to represent the data to be returned to the browser are *JSON* and not XML – contrary to what the name of the paradigm might hint. However, this is not a very personal interpretation of the Ajax paradigm, since the technologies of the web app do not necessarily have to match those listed in its name. Ajax, in fact, merely defines the rules underlying the communication between browser and web server, leaving the developer a certain degree of freedom in terms of the technologies to be used.

7.2.2 Component Architecture and Software Structure

The architecture of the web application is modelled on the *Model-View-Controller* pattern, more often referred to as MVC. As the reader may guess, this architectural pattern allows the logic of our web app to be divided into three interconnected layers, precisely Model, View, and Controller. The underlying principle is once again that of separation of concerns, which not only improves readability, maintainability, and testability of the web app software, but also simplifies its development. The main advantage of this architectural pattern is the sharp separation it ensures between the app data and its presentation, which facilitates the creation of multiple views from the same model; this allows our web app to display information on screen tailored both to the user interacting with it and to the service to be provided. It becomes even possible to separate the processing of inputs and outputs, splitting their responsibilities between controllers and views; without the MVC pattern, in fact, both logics would be placed in the same module, since our web app is used as both input and output interface, with a drastic deterioration in the software quality. The layout of the MVC architecture of our web app is presented in the component diagram of Figure 7.19, which provides, however, a simplified view of the structure of the software; a number of its modules, as well as the relationships between the classes of the various levels, have in fact been omitted for reasons of legibility.

The lowest level of the web app architecture is the *Model*, which is responsible for the application data and its logic. This layer provides a class for each of the *entities* involved in the functionalities of the web app; these are *Measurement*, *Dongle*, *Vehicle*, *Emission Model*, *Staff*, *Staff Role*, *Shift*, *Call*, and *Call Type*, which are modelled on the corresponding database tables. It also includes classes that do not refer to any tables in the database, which are *Operation* and *Vehicle to Dispatch*: these are clusters of information from other entities and external data sources that are only meaningful in the context of the web application. Operation, which is used when monitoring the operational status of fleet vehicles, associates the GPS position of each ambulance with an indication of its availability status and the number of operators that are active on it, information that can be obtained by combining vehicle, call, and shift records; Vehicle to Dispatch, used to keep track of the ambulances that prove suitable to handle a particular emergency, combines the data of each vehicle with the details of the staff operating it, as well as information on the call to be served; each ambulance is also provided with an estimate of the time it requires to reach the destination, information that is not available in the database, but obtainable from external data sources.

The retrieval of data within the Model layer is delegated to a group of classes whose purpose is to simplify the interaction between the web app and the outside world. One of these is *HTTPCommunication*, which allows the web app to retrieve information from external APIs. Among the services the web app interacts with are Address Geocoding and DistanceMatrix, both provided by Google Maps. The former, which converts an address into geographic coordinates, is used to determine the GPS position of the destination based on the location provided by the caller. The latter, which produces distance and travel time between lists of origin-destination points, is employed to identify which vehicle is suitable to serve a call that can reach the destination faster. It should be noted, however, that HTTPCommunication is not meant for database interactions. In fact, unlike the mobile app, which communicates with the data storage via the API, the web app does not rely on mediators to retrieve records from the database: data is extracted by the entity classes themselves through HQL (Hibernate Query Language) queries. The reason behind the usage of this SQL-like language stems from our decision to resort to *Hibernate*, an object-relational mapping tool that helps us handle the mismatch that inevitably occurs between the data types of the Java language and those of a relational database. Since our database benefits from the PostGIS extension, the web app also relies on *Hibernate Spatial*, which provides the SQL dialect for dealing with geographic data in a standardised way.

The Model class responsible for the interactions of the web application with the other components of the platform is, instead, *MQTTCommunication*. This class, which handles the outgoing communication to the MQTT broker, is used downstream of the ambulance dispatching procedure to share the details of the emergency to be served with the operators who are active on the chosen vehicle.

The highest level of the web app architecture is the *View*, which, as its name suggests, consists of software modules responsible for interacting with the user and displaying data on screen. Among the classes of this layer is *Map*, which manages the Google Map on which the web application visually summarises information about the fleet. This level also includes a pair of classes in charge of requesting the information to be presented to the user, which establish periodic communication with the back-end and retrieve the desired data from the database and other sources; the web app relies on a *Multi-Vehicle Polling* class, which is used to retrieve essential information on all the vehicle in a certain area, and on a *Single-Vehicle Polling* class, used to obtain detailed information on a specific ambulance using its ID. The responsibility for reproducing this content on screen is shared between a number of software modules. For example, the web app software is equipped with a *Marker* class that positions and moves the icons on the map, as well as changing their colour. Finally, this level includes a different module for each of the visual components deployed on the dashboard, namely *Form*, *Table*, and *Chart*. Further details on these elements can be found in Section 7.2.3, which describes the design of the web app's UI.

The middle layer of the MVC architecture is the *Controller*, whose responsibility is to retrieve, process, and provide the data to be shown to the user; this layer thus contains the core application logic of the UI and acts as a link between the View and the Model. The components behind the Controller of our web app are the *Jakarta Servlets*, which have already been introduced in Section 5.2.2: upon receiving an HTTP request from the View, they generate a response in JSON containing the Model data, suitably processed, to be shared with the browser. Each servlet is associated with a specific operation of the web application, which can be requested from the front-end by specifying the Uniform Resource Locator (URL), colloquially termed as a web address, that is assigned to it.

As shown in Figure 7.19, the servlet classes are divided into three main packages, according to the services of the web app they are used for. As might be expected, the Operations Monitoring package consists of two servlets: one is used for multi-vehicle monitoring, the other one for single-vehicle monitoring; the same applies to the diagnostics monitoring service: the two servlets, in fact, despite operating on different datasets, perform essentially the same function. The vehicle dispatch procedure requires, instead, three different servlets: one is responsible for creating the call in the database based on the information provided by the call handler, one for selecting the ambulance to be activated based on the status of the fleet, and one for assigning the selected vehicle to the call and notifying its staff. The structure of this package reveals our decision to implement a *supervised dispatching service*, which prompts the call handler to confirm the activation of the chosen ambulance or to select an alternative one; however, nothing would prevent us from combining the functionalities of the three servlets and achieving fully automated dispatching, in which the vehicle is automatically activated following the creation of the call. A servlet for monitoring the dispatched vehicle is also provided, whose functioning, together with that of the other servlets, is discussed in the pages that follow.

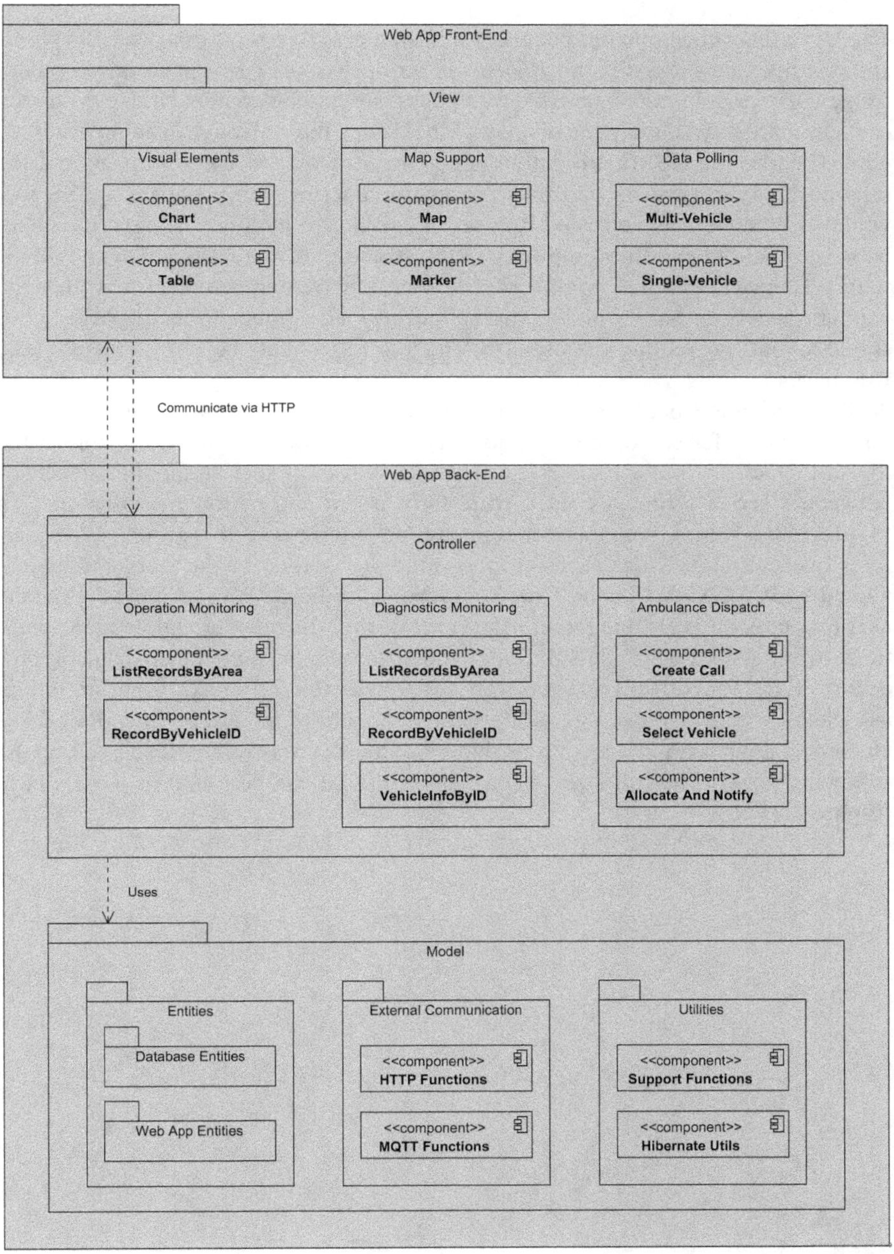

FIGURE 7.19 Web app, software architecture.

This component diagram offers an oversimplified view of the software structure, since it omits several modules and all the dependencies between the included elements.

7.2.3 UI Design and Component Functionalities

The UI of the web application consists of a single web page; it is written in HTML and has been developed using *Bootstrap*, an open-source front-end development framework for creating responsive web pages, so that the content shown on screen can automatically adapt to the display of the device the web page is being accessed with. Clearly, the HTML file only defines the structure of the web page and the relationships between its parts; its presentation style is entrusted to a CSS file, while its behaviour is entrusted to a set of JavaScript modules. It must be noted, however, that the extreme simplicity in the structure of the UI is not synonymous with a lack of functional complexity; it should be remembered, in fact, that the web app has been built with Ajax, which allows all content to be displayed on a single layout, preventing the user from having to navigate between several static web pages.

The navigation between the various sections of the web page can take place through two different components, namely a *navigation bar*, always visible at the top of the screen, and a *carousel*, which is a slideshow that automatically scrolls between a series of images, each with its own text and buttons, as illustrated in Figure 7.20. Overall, both these components are equipped with a group of four *buttons*: one button provides access to the value proposition of the web application, named VEICOLNET, like the other components of the CVP, two buttons allow the user to access the real-time monitoring functions for diagnostics and vehicle operations, while a final button lets the call handler initiate the ambulance dispatch procedure. Each functionality of the web app, retrievable by clicking on the button associated with it, is linked to a different component of the UI; the layout of these elements, along with the operations underpinning them, is briefly discussed on the following pages. The full-size screenshots of the UI can be found in Appendix D (Figure 7.21).

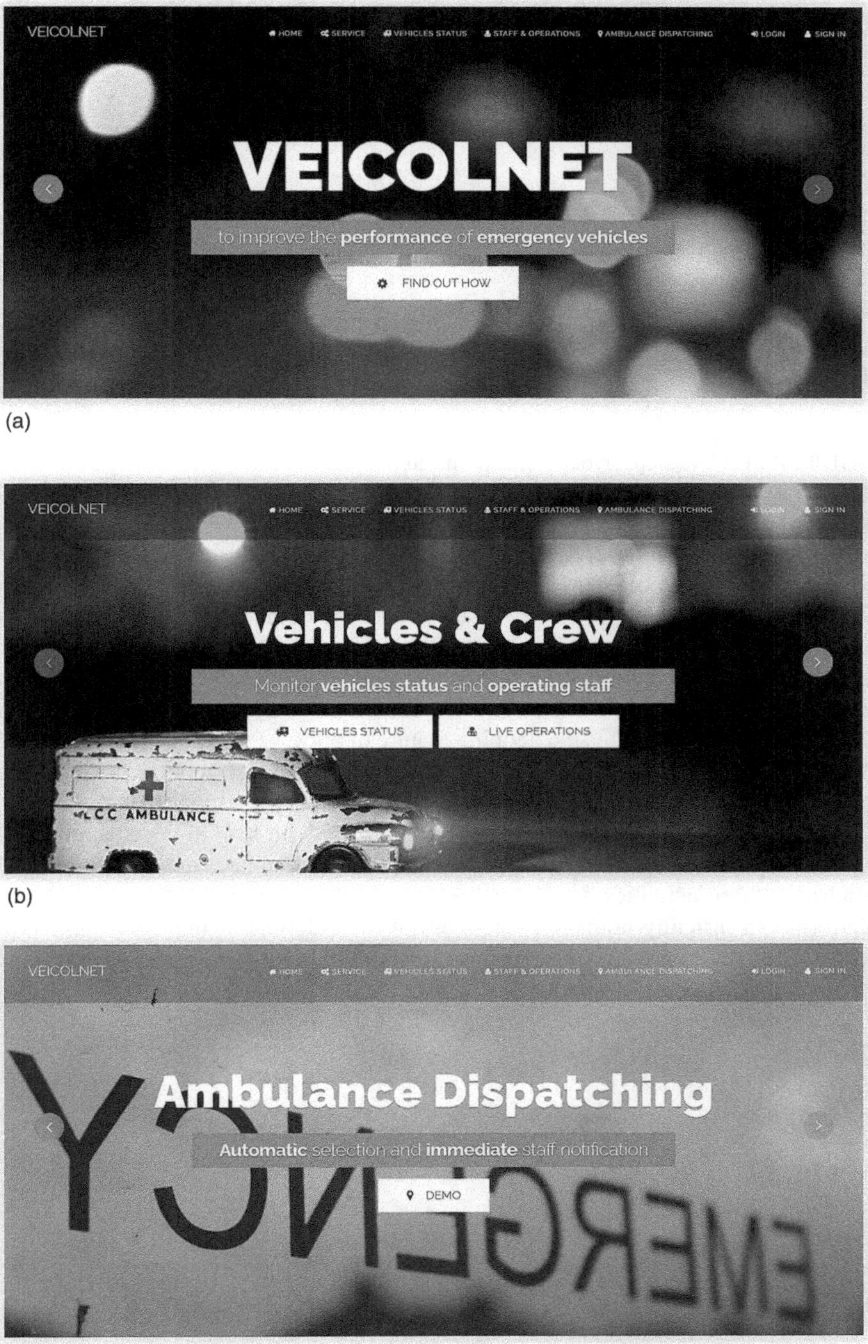

FIGURE 7.20 Web app, layout: Bootstrap carousel. (a) Measurement data transmission and storage. (b) Component notification upon vehicle dispatch. (c) Component notification upon call closing.

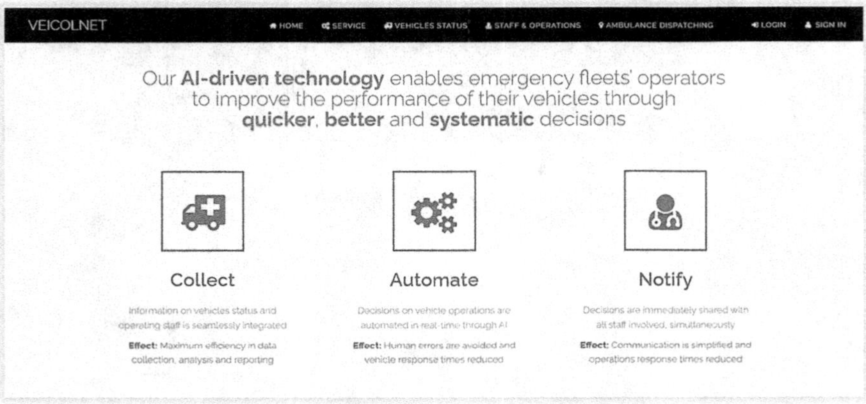

FIGURE 7.21 Web app, layout: Value proposition.

Service is the section of the web page that contains the value proposition of our data platform. It is located immediately below the carousel and can be accessed by clicking the Service button of the navigation bar, by clicking the button included in the first slide of the carousel, or by simply scrolling down the web page. This screen summarises the Unique Selling Points (USPs) of the CVP, namely the advantages that our platform provides over commercial systems for emergency vehicle management. As highlighted by the first USP, the defining feature of our CVP lies in its ability to seamlessly integrate vehicle data context data, which allows the web application to perform queries on a dataset that is not made available by any other existing system. Access to this comprehensive set of data provides our platform with several benefits, including the possibility of automating decision-making, which eliminates the need for call dispatchers, whose activities are invariably exposed to errors and delays. Finally, our platform can instantly notify the operators of the outcome of this decision-making process, further improving the response times to vehicle operations. The reader is referred to Chapter 8 for evidence supporting the above claims on the USPs.

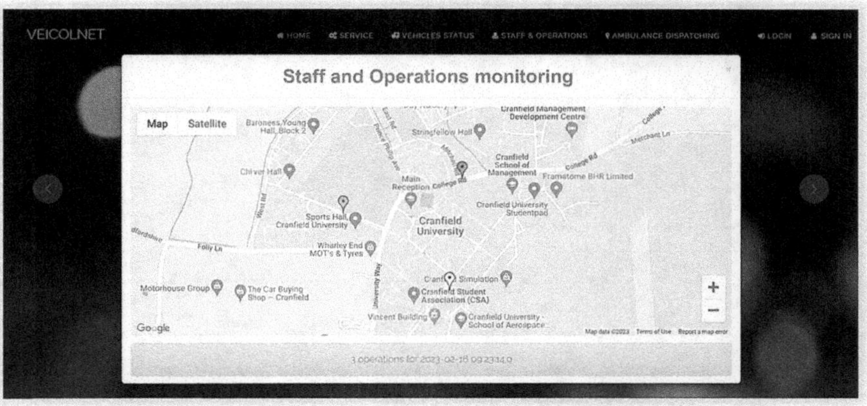

FIGURE 7.22 Web app, layout: Multi-vehicle monitoring.

Multi-Vehicle Modal is the element of the UI associated with the multi-vehicle monitoring function, which is shown on screen when the user clicks on one of the buttons related to fleet monitoring. Figure 7.22 illustrates the modal for monitoring fleet operations, whose layout is identical to that of the diagnostics monitoring function, with the exception of the heading. Even the operations behind the two modals are equivalent in many respects: in both cases, every time the Google Map is moved or zoomed, the web app obtains its boundaries, which are used to request the latest data for the ambulances in that geographic area, with a polling frequency of 1 Hz; what differentiates the two functions is the dataset retrieved from the database, as evidenced by the queries in Appendix D. The records obtained are used to place and move markers on the map, as well as changing their colour. Each marker is also provided with an event listener, which allows the user to monitor a particular ambulance by clicking on the icon associated with it. This listener, however, is disabled if a vehicle does not produce any measurements for more than 10 seconds: this is a time window used to take into account the natural delays on the network before assuming that a dongle is inoperative.

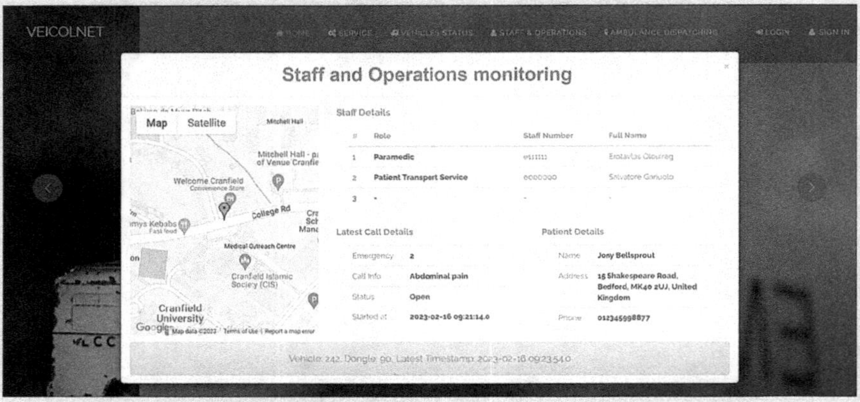

FIGURE 7.23 Web app, layout: Single-vehicle operations monitoring.

Single-Vehicle Modal, Operations is the dialog box used to display information on the operational status of a particular ambulance; it is shown on screen when the user clicks on one of the markers of the corresponding multi-vehicle monitoring function, which leads to the removal of all other icons on the map. Following this interaction, the web app modifies the map settings, disabling any user interaction with it, to ensure that the screen always stays centred on the selected marker. This is the same modal used for multi-vehicle monitoring, whose layout is modified to make room for a group of three bootstrap *tables*, namely staff, operation, and patient. Once the layout has been altered, the web app starts to retrieve the data of the chosen vehicle with a frequency of 1 Hz. The records obtained are used to populate these tables, change the position of the marker and centre the map on it, and determine the colour of this icon. The possible colours are *green*, signifying a vehicle with crew that is currently available, *yellow*, for a vehicle that is free but unmanned, and *red*, for a vehicle that is not available since it is engaged in a call; *violet*, whose meaning is common to all forms of monitoring, denotes an ambulance whose dongle has not provided any data for more than 10 seconds.

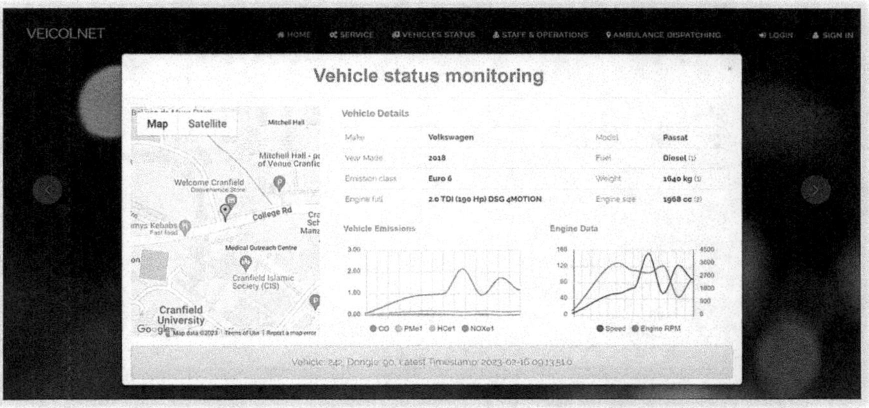

FIGURE 7.24 Web app, layout: Single-vehicle diagnostics monitoring.

Single-Vehicle Modal, Diagnostics is the dialog box used to display information on the diagnostics of a certain ambulance, which is shown on screen when the user clicks on one of the icons of the corresponding multi-vehicle monitoring function. This version of the modal is equipped with a single bootstrap *table*, used to display the technical specs of the ambulance, and two *line charts* built with *ApexCharts* – a JavaScript library for developing interactive graphs for the web. One chart is used to profile the speed of the ambulance and that of its engine in real-time, the other to display a live estimate of the emissions it produces. This form of single-vehicle monitoring involves two data requests to the back-end: a one-off query to retrieve vehicle specifications, which – being static data – would be superfluous to request periodically, and a recurring request with a frequency of 1 Hz, employed to obtain the most recent measurement within the last 10 seconds. The retrieved records are used to populate the table, to move the marker and centre the map on it, and to set the icon colour; the colours are the same as those used for operation monitoring, albeit with a different meaning: while *cyan* stands for a vehicle with the engine off, *green*, *yellow*, and *red* denote increasing amounts of CO emitted.

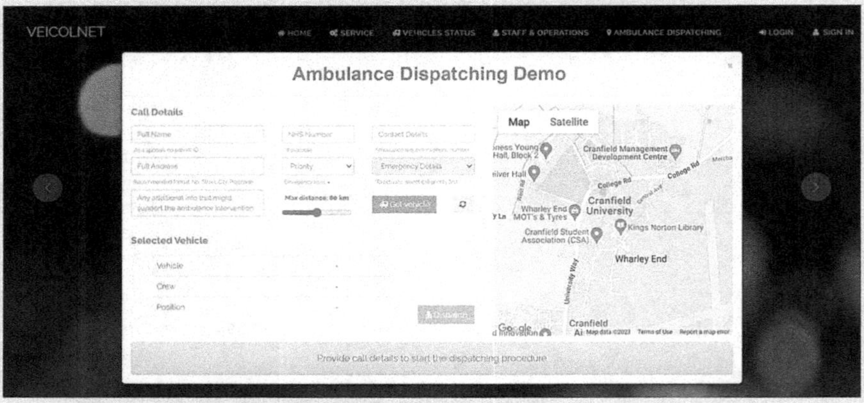

FIGURE 7.25 Web app, layout: Ambulance dispatch, empty form.

Ambulance Dispatch Form is the modal used to control the ambulance dispatch procedure and to monitor the vehicle that is activated through it; it is displayed on screen when the user clicks either the corresponding button on the navbar or the button embedded in the third slide of the carousel. This element of the UI consists of three sections. The first, located in the top left-hand corner, contains a *form* that mirrors the structure of the call table in the data storage, allowing the call handler to provide all the details of an emergency. As mentioned in Section 7.2.2, some fields of this form leverage the asynchronous communication enabled by Ajax. Full Address, for example, is linked to the *Address Autocomplete* service by Geoapify, which automatically predicts the addresses during user-entry, whose suggestions are displayed in a list that unfolds below the field without the need to refresh the entire page; recourse to this API is of utmost importance as it speeds up data input and it ensures that only verified and standardised addresses are stored in the database. The other field to take advantage of Ajax is *Emergency Details*: this drop-down list is dynamically generated to solely display the emergencies related to the chosen emergency level, further simplifying the activity of the call handler.

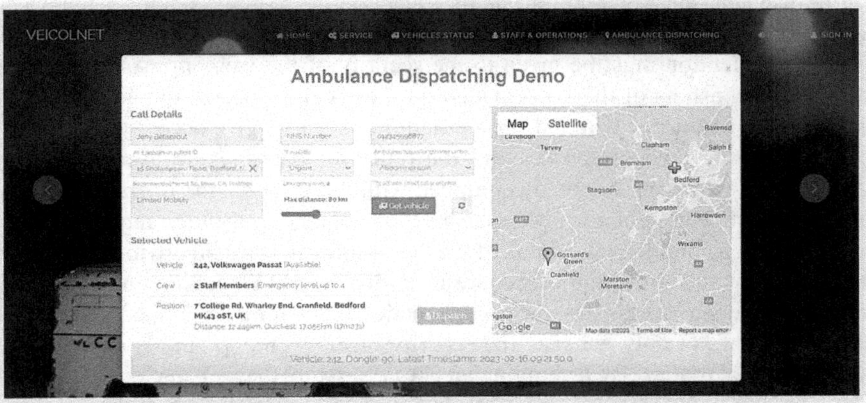

FIGURE 7.26 Web app, layout: Ambulance dispatch, selected vehicle.

Once all the mandatory information about the emergency has been entered, clicking on the *Get Vehicle* button instructs the web app to save the call in the database, to freeze the form fields so that they cannot be edited unless the entire dispatching procedure is restarted, and to initiate the vehicle selection process. The selection of the ambulance to be activated is carried out in two stages: the first stage selects, within the maximum distance provided, the five available vehicles closest to the destination that are able to clinically handle the emergency; this operation, based on database data alone, relies on a spatial analysis with PostGIS. The second stage resorts to DistanceMatrix to determine which of these ambulances is the fastest to reach the destination, based on live traffic conditions. Recourse to a preliminary screening aims to reduce the number of items to be submitted to the external API, which ensures a reduction in the cost of this third-party service. This, however, could reduce the quality of the dispatching process, as there is a trade-off between the precision of the selection procedure and the cost of DistanceMatrix. Indeed, as the number of ambulances submitted to Google Maps increases – and thus the cost of the API call – the likelihood of selecting the most suitable vehicle also rises, since live traffic information might significantly alter the choice based on distance alone.

Following the selection of the ambulance, the web app places both the destination icon and the chosen vehicle's icon on the *map*, located to the right of the form, and centres the map to display both. The information on the selected ambulance is presented in a bootstrap *table*, located in the lower left section of the modal; the table includes the location of the ambulance together with an estimate of the time required to reach the destination, which is updated every 5 seconds. The selection of the ambulance also enables the *Dispatch* button, which allows the call handler to activate the chosen vehicle: interaction with this button instructs the web app to store the outcome of the dispatch procedure in the database and to notify the activation of the vehicle to its staff via MQTT, once the system has verified that the vehicle still meets the requirements to serve the call. If the call handler reuses the *Get Vehicle* button before dispatch, the web app repeats the selection procedure; clicking the adjacent button empties the form, instead.

Finally, attention must be given to the security of the web app. As can be seen from the screenshots of the UI provided on the previous pages, the navigation bar includes two additional buttons: *sign-in* and *login*. These give access to two *forms*, one for creating the user profile and the other for entering the user credentials. Although the web page is already equipped with these forms, the operations behind them have not been implemented, as security is not among the core requirements of our design. In a commercial version of the data platform, it would obviously be desirable to use an authentication procedure to initiate an HTTP session, allowing only authorised users to access the web app's functionalities for a predefined period. This strategy would not only enhance the security of the web app but also benefit fleet management activities. By leveraging an HTTP session, the web app could track the user interacting with the data platform and store in the database the identity of the call handler responsible for dispatching a particular ambulance. Implementing this would, of course, require minor modifications to the CVP, such as altering the database structure to create an association between the call and staff tables. In conclusion, for the commercial version of the CVP, our recommendation is to implement state-of-the-art security strategies, such as multi-factor authentication, for details of which the reader is referred to external sources.

7.3 CHAPTER SUMMARY

This chapter has examined the design of the user interfaces that a CVP must adopt to guarantee an effective management of the ambulance fleet of the NHS. The two interfaces are by no means innovative from a purely technological perspective, as every fleet management system normally uses either a mobile application or a web application. This chapter has demonstrated, however, that these user interfaces are pioneering from a functional standpoint; the innovation is the result of the design choices made for the other components of the data platform, which enable functionalities that are currently unattainable with any other commercial product for emergency fleet management. Having described the design of all components of the data platform and obtained a prototype implementation of the entire system, Chapter 8 can demonstrate the ability of the CVP to meet the unusual requirements of the context-aware, real-time fleet management applications.

8 Data Platform Testing

Software testing is the process of verifying and validating the behaviour of a software product [95]. Software testing activities can be used to determine whether individual components of a system function correctly on their own or work together as intended; they can also be employed to ensure that the overall system is free of vulnerabilities, that it properly handles a certain workload, and that it meets the needs and expectations of its users. A wide variety of software testing activities is clearly available, and the choice of the one to adopt largely depends on the testing objective that is pursued.

The aim of this chapter is to demonstrate that the data platform developed as part of this research is able to meet the requirements of a context-aware, real-time fleet management application, enabling performance and functionalities that cannot be achieved with the existing systems. To this end, we employ functional testing, which allows us to assess the behaviour of the fleet management activities against a real-world scenario, treating the underlying platform as a black-box component. The procedure we focus on is that of vehicle dispatching; among the various fleet management activities, it is the one that requires both context-awareness – having to combine vehicle measurements and context data – and real-time responsiveness, as the outcome must be obtained within a certain time frame to be considered correct. The chapter opens with a section on "Testbed Description", which describes the characteristics of the testbed and explains our decision not to resort to performance testing. The next section presents the functional tests used to demonstrate the platform's ability to automate the decision-making behind ambulance dispatch, eliminating both errors and delays that are caused by human operators. The final section, which validates the test results, closes the chapter.

8.1 TESTBED DESCRIPTION

The testbed used to assess the behaviour of the ambulance dispatching procedure is a working prototype of the data platform outlined earlier in the book. It includes most of the hardware, software, and networking elements envisioned in the CVP design, though some components may differ slightly from those that would be featured in the final version of the platform. The testbed should, in fact, be seen as a simplified replica of the CVP. This, however, does not compromise the reliability of the tests. On the contrary, it allows the desired analyses to be performed without having to resort to the definitive version of the CVP, which would be unnecessary for the purposes of this research.

DOI: 10.1201/9781003688549-8

Being a simplified version of the data platform described in this book, the testbed's structure can be directly inferred from the previous chapters, by supplementing the blueprint of the CVP with details on the changes made to its components. For instance, the testbed's database relies solely on time partitioning, as the entire dataset is located on a single node, making any form of spatial partitioning unnecessary. In addition, the database server runs on a local machine – a Lenovo ThinkPad X1 Carbon Gen 4 laptop, equipped with an Intel(R) Core(TM) i5-6300U CPU at 2.50 GHz and 8 GB of RAM – rather than on a cloud platform. Regarding the communication infrastructure, the testbed employs both HTTP and MQTT, but forgoes the use of Kafka, as the current version of the dispatch procedure does not include any event-driven notification mechanism. HTTP communication takes place via an API that is running on the local machine, which can be accessed with ngrok, a paid service that exposes local servers behind NATs or firewalls to the public Internet via a secure tunnel. MQTT communication, on the other hand, occurs via a public HiveMQ broker. The OBD dongles used in the testbed maintain the same structural and functional characteristics described in the vehicle interface chapter; the only notable difference lies in the positioning software, which overwrites the actual position of these devices with a fictitious location, allowing us to obtain the fleet configuration required for testing without having to physically move the vehicles. The user interfaces of the testbed, instead, do not deviate in the slightest from the components presented in this book. The mobile app runs on a Xiaomi Redmi Note 10 Pro, both on the physical device and in a simulated environment via the Android Studio emulator. The web app, also executed locally, has been deployed to Apache Tomcat 8.5, which combines the functionalities of a web server and those of a servlet container.

Our decision not to resort to non-functional testing stems precisely from the differences that arise between the characteristics of the testbed and those of a hypothetical final version of the data platform. These differences do not influence the outcome of a functional test, but would be a determining factor in analyses that focus on the performance of the CVP, such as a reliability or stress test. Indeed, such tests would be meaningful only if a near-market version of the data platform were used – one that not only adheres to the design described in this book, but also incorporates design considerations related to system performance that are required for a commercial product, which are beyond the scope of this project. For example, the database tables would need proper indexing, and queries would need optimisation; it might also be advisable to implement a master-slave strategy to optimise database I/O and take measures to improve API reliability. These are just a few of the interventions that might be required to satisfy all the non-functional requirements of a commercial system.

8.2 FUNCTIONAL TESTING

The notion of *functional testing* refers to the quality assurance process that seeks to establish whether a software component conforms to its functional requirements. A functional test is conducted by providing a specific input to the element under test and evaluating the output it returns, allowing one to determine whether its behaviour aligns with what is expected of it. It is, by definition, a *black-box test*, as no specific knowledge of the application code or the internal structure of the component is required.

This research resorts to functional testing to demonstrate that the CVP described in this book enables the automation of the decision-making process behind the ambulance dispatch, eliminating both the errors and delays that are inevitably caused by call dispatchers – particularly when operating under high-pressure conditions. A positive test result would thus support the central hypothesis of this work: improving the operational productivity of the NHS ambulance service does not necessarily require an increase in its workforce, but rather the use of a bespoke data platform that enhances its data management. The testing carried out within this project, however, is limited to evaluating the benefits of the CVP on the vehicle dispatching procedure itself; the potential impact that the situation data recorded during dispatching has on other fleet management activities is therefore not considered. Such data could be used, for example, to tailor the rotas of ambulance operators to call demand or to determine a more efficient deployment of the ambulance fleet across the territory, further boosting the performance of the control centres of the NHS.

As can be inferred from the description of the testbed, the functional tests employed in this project are *system tests*, as they involve all the elements that comprise the data platform. Consequently, this testing activity is carried out downstream of the unit and integration tests already performed on the various components of the CVP; these earlier tests are not included in the book, as they do not provide insights relevant to the purposes of this research. The tests conducted to assess the platform's functionalities fall into the category of *scenario-based testing*, which relies on sequences of actions and events that reflect typical system usage. This is a well-established approach in the automotive industry, where it is often employed to assess the safety of automated vehicles, as it reduces the infinite possibilities of behaviour of a highly complex system to a finite set of representative scenarios, making the testing process more manageable.

Our testing activity involves a single test scenario consisting of three test cases, each based on a specific fleet configuration and a distinct emergency situation. These tests are presented in the following sections, each of which offers a brief description of the test, information on its inputs, preconditions, and postconditions, and its result. Each test is designed to highlight a particular advantage of our platform over the existing dispatching systems. The objectives of these tests are summarised below:

Functional Test 1 aims to demonstrate that the unique set of technologies chosen to implement the CVP enables automatic and quasi-immediate ambulance dispatching, a capability not achievable with current fleet management solutions. It should be noted that the procedure under test includes not only ambulance selection, but also all the manual activities that are currently handled by call dispatchers, from communicating the outcome of the procedure to updating the status of the call and that of the vehicle.

Functional Test 2 aims to demonstrate that the integration of vehicle and context data relieves call dispatchers of the burden of selecting the appropriate ambulance for a given emergency, eliminating a manual process that is inherently prone to errors.

Functional Test 3 aims to demonstrate that the aggregation of situation data stored in the CVP with live information from other sources – such as traffic data provided by Google Maps – enables a higher quality decision-making process than the one that is currently achievable in the control centres of the NHS.

8.2.1 Role of a Bespoke CVP in Automating Decision-Making

TABLE 8.1
Functional Test 1: Testcase Description

Test Case Description	
Title	Automation of the ambulance dispatch procedure
Type	Functional
Description	It verifies that the CVP is able to automate (i) the creation of a call record in the database, (ii) the selection of a suitable vehicle for the emergency and, once the vehicle is activated, (iii) the updating of the call status in the database, and (iv) the transmission of its details to the crew operating it.
Users Involved	Call Handler, Ambulance Operators (Test1, Test2).
Preconditions	• Vehicle 242, equipped with a dongle, must have been placed at Bedford Hospital, A&E Department (52.128, -0.471); the vehicle must not be active on any emergency call; • Test1 (UCA, Level 2) must have started a shift on v. 242; • Test2 (PTS, Level 1) must have started a shift on v. 242.
Trigger and Data	The call handler receives an Urgent call (Level 2) from Cranfield University, Moulsoe Road, Cranfield, MK430AL, UK.
Ordinary Sequence	• The call handler enters the emergency information into the dispatch form of the web application; • The call handler presses the *Get Vehicle* button of the form; • The call handler activates the selected vehicle by pressing the *Dispatch* button of the form.
Expected Outcome	The call handler receives on-screen confirmation that vehicle 242 has been correctly activated on the current call.
Postconditions	• A new call record has been created in the database based on the data provided by the call handler; this record is linked to that of vehicle 242 via a foreign key; • The web app displays an estimate of the time required for vehicle 242 to reach the call destination; • Ambulance operators Test1 and Test2 simultaneously receive a notification on their mobile app that vehicle 242 has been activated on the new call, together with call details.
Status	Pass

FIGURE 8.1 Functional Test 1: Preconditions (vehicle and staff details).

This screenshot of the web application UI provides a visual representation of the test preconditions, namely the location of the single vehicle involved, its operational status, and details of the ambulance operators that are currently active on it.

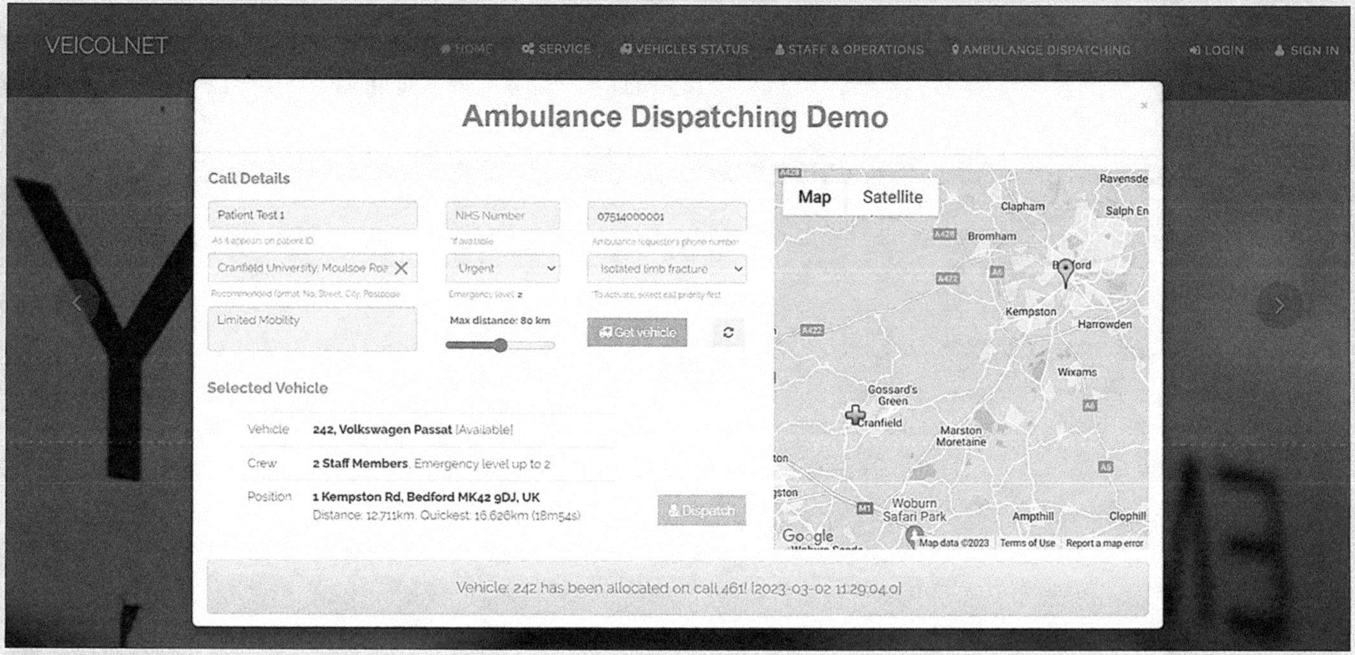

FIGURE 8.2 Functional Test 1: Data input and test result.

This screenshot of the web application UI provides a visual representation of the data used to simulate the emergency; it also shows the outcome of the dispatch procedure, proving the CVP's ability to associate the intended vehicle with the newly generated call.

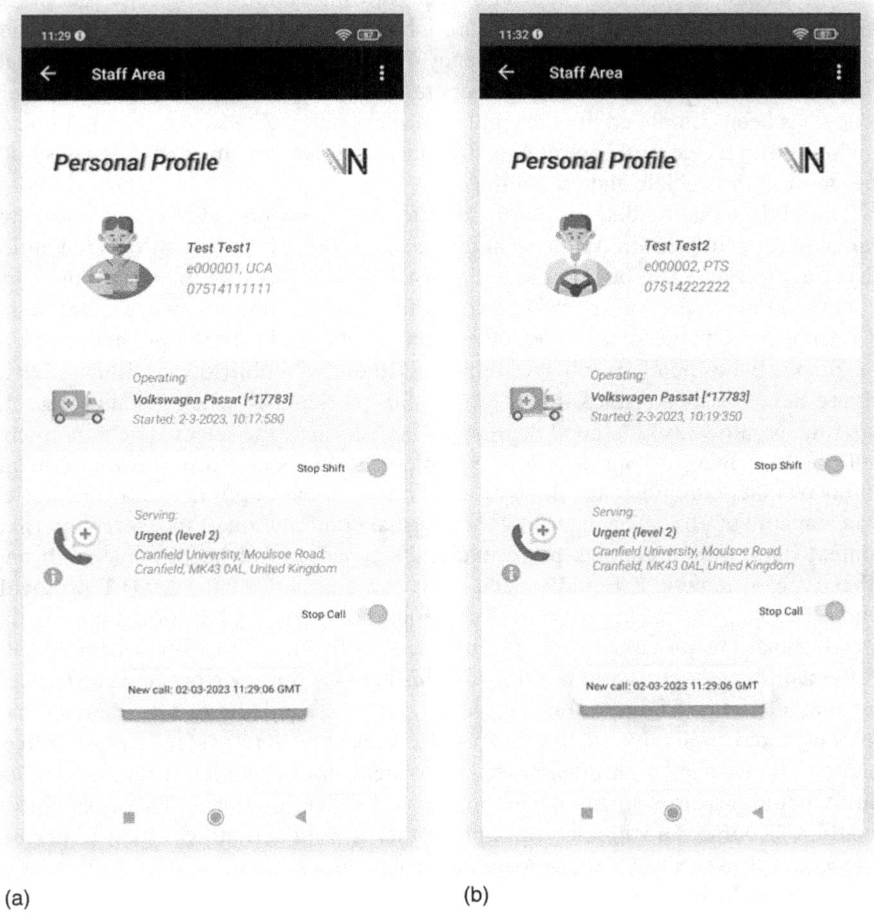

FIGURE 8.3 Functional Test 1: Postconditions (staff notification). (a) Test1, personal page. (b) Test2, personal page.

These screenshots of the mobile app UI prove the CVP's ability to automatically and simultaneously notify ambulance operators of the outcome of the dispatch procedure.

The above figures (Figures 8.1–8.3) attest to the ability of our data platform to automate all tasks involved in the ambulance dispatch procedure – from the creation of a call record in the database to the selection of an appropriate vehicle, the allocation of that vehicle to the call, and the communication of emergency information to the personnel operating it. This outcome is the result of the unique combination of technologies the CVP has been built with, which radically differ from those featured in conventional emergency FMS. Such systems, for example, lack a component for collecting data on ambulance operators, preventing the implementation of an automatic procedure that selects vehicles based on the clinical capabilities of the crews currently on duty (Table 8.1).

Nowadays, even the communication between the control centre and the ambulance operators is handled manually, through the use of radio communication technologies. The transmission of emergency information is carried out by the call dispatcher, who contacts the crew that operates the selected vehicle. Once the emergency has been completed, it is the ambulance operators who use radio communication to inform the control centre, with the call dispatcher who manually updates both the status of the vehicle and that of the call.

The ability of our data platform to automate decision-making has a distinct impact on the mobilisation time of ambulances. The CVP, in fact, allows the entire dispatching process to be completed in just *four seconds* – an interval that spans from the moment the call record is created downstream of triage, whose timestamp is retrieved from the database, to the moment the ambulance operators receive the dispatch notification and the vehicle is thereby mobilised, as illustrated in Figure 8.3. This is a remarkable achievement, particularly when considering that this time window also included the manual activation of the selected vehicle by the call handler, owing to our decision to implement a supervised procedure. Crucial to this result is our choice to employ a relational database optimised for time-series data, capable of efficiently handling the unusual combination of measurements and context data, which enables performances that would not be possible with any other type of database. Equally decisive is the adoption of the MQTT protocol, which ensures the reactive and simultaneous notification of the outcome of the decision-making process to all the intended recipients. However, as previously explained, the reader should not mistake this activity for a performance test, given the characteristics of the testbed employed. Nevertheless, the results obtained can be considered indicative of the performance that the final system would deliver under realistic load conditions, based on the behaviour expected from the individual technologies used in the development of the data platform. This assumption should, of course, be validated through further testing activities – such as performance and stress tests, to be carried out on a near-market version of the CVP under realistic workloads.

The automation of the decision-making process not only accelerates the entire dispatching procedure but also eliminates errors from the vehicle selection process, as demonstrated in the following section. This is a combination of advantages that is inconceivable in the control centres of the NHS. In fact, as the monthly performance reports of the ambulance service reveal, control centres are indeed capable of completing the dispatching procedure with remarkable speed, particularly in the case of a high-acuity call. However, as acknowledged by the ambulance trusts themselves, this task is so complex that it often leaves call dispatchers with little time to assess whether the available ambulances can clinically serve the emergency or reach the destination on time. This inevitably paves the way for errors in the vehicle selection process, with call dispatchers frequently compelled to activate more than one ambulance in an attempt to ensure that the emergency is correctly served within the required time target. Such practices lead to a progressive depletion of available resources, ultimately reducing the ability to handle subsequent emergencies, which slowly deteriorates the overall performance of the ambulance service.

Following on from these findings, what is most surprising is the widely accepted belief that NHS trusts can do little to influence the ambulance responses to emergencies, and that the only valid answer to improving the operational productivity of the ambulance service lies in expanding its workforce. This only reaffirms that it is not the lack of a technological solution that hinders the resolution of the problem, but the inability of the NHS to identify its root causes.

8.2.2 Benefits of Integrating Vehicle Data and Context Data

TABLE 8.2
Functional Test 2: Testcase Description

Test Case Description	
Title	Vehicle selection process with vehicle and context data
Type	Functional
Description	It verifies that the CVP is able to select only ambulances that are (i) active, i.e. equipped with a dongle that has produced valid measurements in the last ten seconds, (ii) available, i.e. not engaged in any call, and (iii) able to clinically serve the emergency, i.e. with staff whose overall skill level is equal to or higher than the emergency level of the call.
Users Involved	Call Handler
Preconditions	• Vehicle 239, equipped with a dongle, must have been placed at 8 Gulliver Cl, Kempston, MK42 8RB (52.122, -0.499); it must have a skill level of 4 and be already serving a call; • Vehicle 240, equipped with a dongle, must have been placed at 8 Addison Cl, Kempston, MK42 8JA (52.117, -0.489); it must have a skill level of 2 and be available; • Vehicle 241, equipped with a dongle, must have been placed at 2 Victoria Rd, Bedford, MK42 9JS (52.128, -0.475); it must have a skill level of 0 (unmanned) and be available; • Vehicle 242, equipped with a dongle, must have been placed at 1 Kempston Rd, Bedford, MK42 9DJ (52.128, -0.472); its dongle must be removed once the vehicle is positioned; • Vehicle 244, equipped with a dongle, must have been placed at 26 Goldington Rd, Bedford, MK403DY (52.140,-0.455); it must have a skill level of 4 and be available.
Trigger and Data	The call handler receives a life-threatening call (Level 4) from the Kempston Hammers Sport Club, MK42 7BN, UK.
Ordinary Sequence	• The call handler enters the emergency data in the form; • The call handler presses the *Get Vehicle* button of the form.
Expected Outcome	The call handler receives on-screen confirmation that vehicle 244 is best suited to serve the current call.
Postconditions	–
Status	Pass

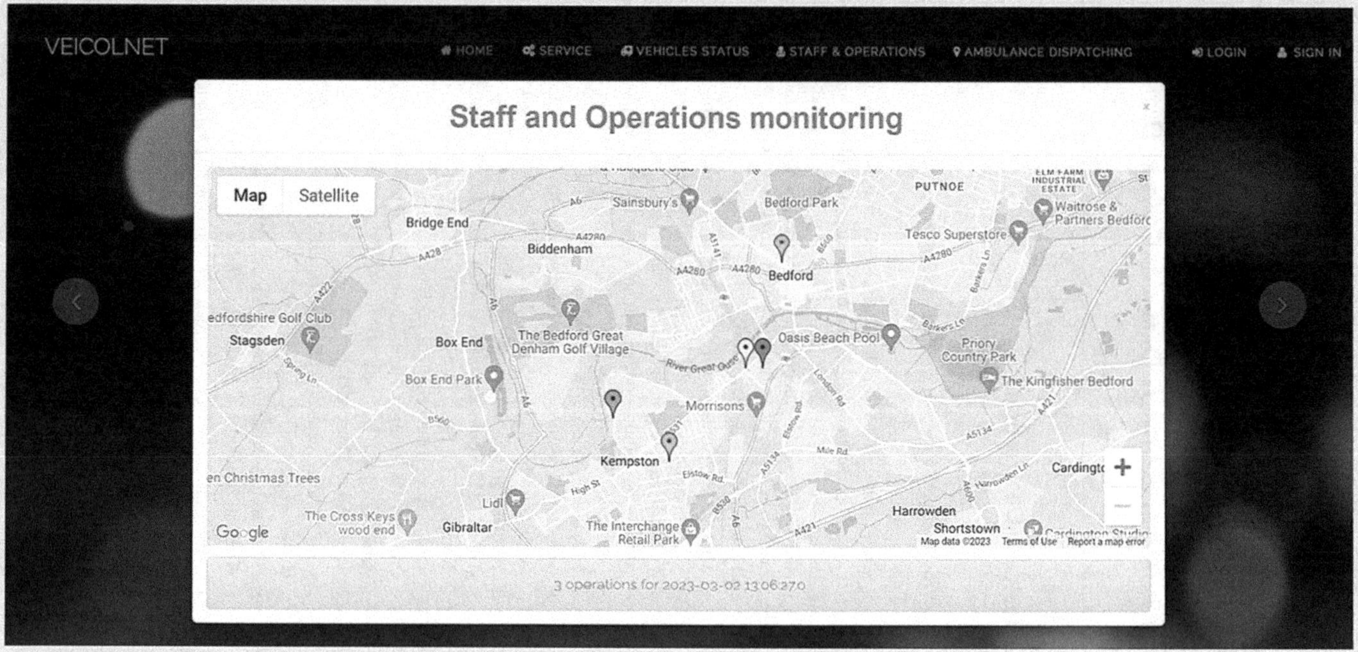

FIGURE 8.4 Functional Test 2: Preconditions (fleet location and status).

This screenshot of the web application UI provides a visual representation of the fleet used for testing; it gives an overview of the geographical layout of the vehicles and their operational status, indicated by the colour of their markers.

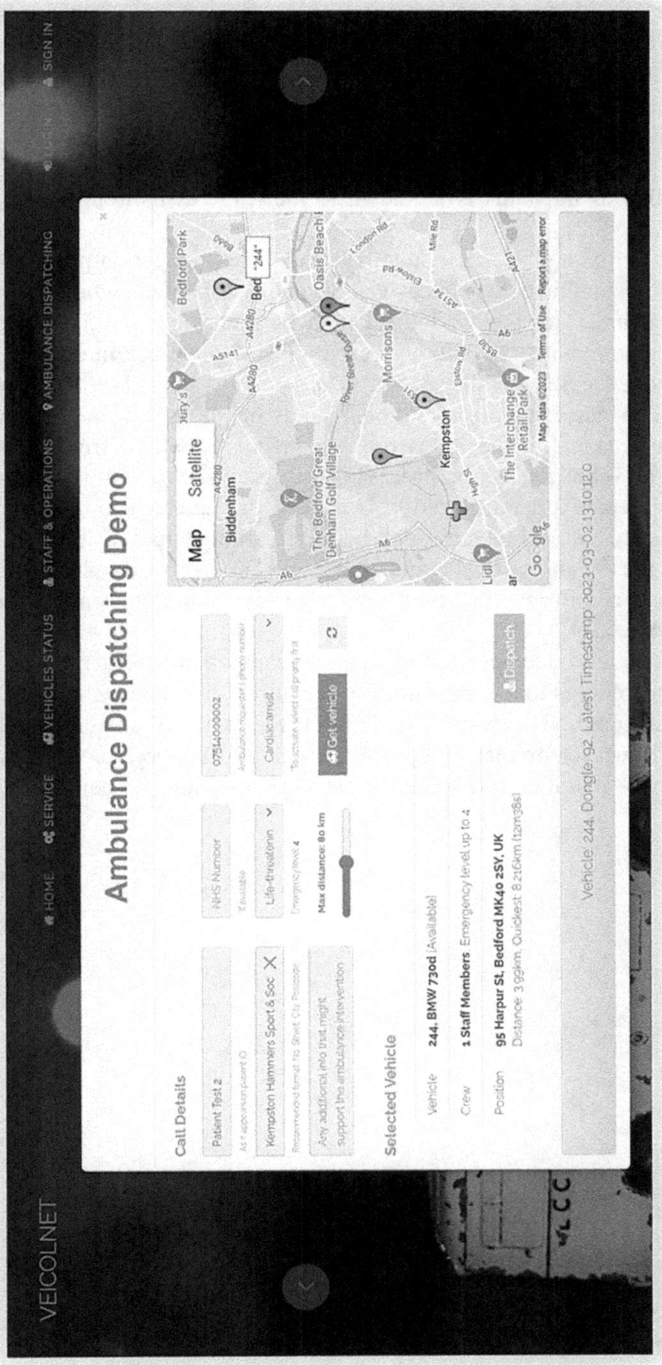

FIGURE 8.5 Functional Test 2: Data input and test result.

This screenshot of the web application UI shows the data used to simulate the emergency and the outcome of the selection process; the functionality of the web app has been modified to display the location of the entire fleet in relation to the destination of the call.

This test, as evidenced by the images above (Figures 8.4 and 8.5), demonstrates the ability of our data platform to select only vehicles that prove suitable to handle the call to be served, regardless of their distance from the destination. The vehicle selection algorithm – whose details can be found in Appendix D – excludes, in fact, vehicle 239, since it is already serving a call, and vehicle 241, which is temporarily unmanned. It also filters out vehicle 240, which, although available and manned, lacks the overall skill level to clinically manage the emergency. Vehicle 242, by contrast, is disregarded as its status is unknown to the CVP, as its dongle no longer transmits data (Table 8.2).

The ability of our data platform to automate the vehicle selection process stems from its capacity to simultaneously query vehicle and context data, which are stored in the same storage unit. This is a result that cannot be achieved in the control centres of the NHS, where context data – although used – is not integrated into the existing FMS. As a consequence, the call dispatcher must manually assess both the capabilities of the crews on duty and the operational status of the ambulances, through a process that, being manual, is inevitably prone to errors. Equally detrimental to the effectiveness of the vehicle selection process is the absence of an automated procedure for collecting context data. Indeed, it is the call dispatcher who is responsible for updating the status of each vehicle and the information about the staff. This further undermines the reliability of the ambulance selection process, as any error or delay in updating this information would result in a dataset that does not reflect the actual state of the fleet.

The result of this test becomes even more significant when considering the number of standard procedures that control staff are required to follow, which can reach as many as 170 in some NHS trusts. This figure is indicative of a decision-making process that has grown so complex and time-consuming – and therefore increasingly susceptible to error – that it can no longer be entrusted to a human operator.

8.2.3 Benefits of Aggregating Data from External Sources

TABLE 8.3
Functional Test 3: Testcase Description

Test Case Description	
Title	Vehicle selection process with driving data from an external API
Type	Functional
Description	It verifies that the CVP is able to select the ambulance with the shortest driving time to the destination from those that are suitable to serve the call, regardless of the linear distance between the vehicles and the location of the emergency.
Users Involved	Call Handler
Preconditions	• Vehicle 240, equipped with a dongle, must have been placed at 17 Colesden Rd, Wilden, MK44 2QL (52.185, -0.376); it must have a skill level of 4 and be available; • Vehicle 241, equipped with a dongle, must have been placed at 1 Church End, Ravensden, MK44 2DA (52.117, -0.425); it must have a skill level of 4 and be available; • Vehicle 242, equipped with a dongle, must have been placed at 26 Home Cl, Renhold, MK41 0LB (52.162, -0.425); it must have a skill level of 4 and be available; • Traffic closure for planned roadworks on Wilden Road, Renhold, MK41 0, from house number 16 (52.162, -0.421) to house number 34 (52.163, -0.418).
Trigger and Data	The call handler receives a life-threatening call (Level 4) from Renhold Village Hall, MK41 0JR, UK.
Ordinary Sequence	• The call handler enters the emergency data in the form; • The call handler presses the *Get Vehicle* button of the form.
Expected Outcome	The call handler receives on-screen confirmation that vehicle 240 is best suited to serve the current call.
Postconditions	The website receives a JSON message from the web server containing the ID of the chosen ambulance; it also includes test-related information such as the linear distance between the vehicles and the call destination, and an estimate of the driving time required by each vehicle to reach the emergency, obtained from Google Maps API based on live traffic data.
Status	Pass

FIGURE 8.6 Functional Test 3: Preconditions (fleet location and status).

This screenshot of the web application UI provides a visual representation of the fleet used for testing; it gives an overview of the geographical layout of the vehicles and their operational status, indicated by the colour of their markers.

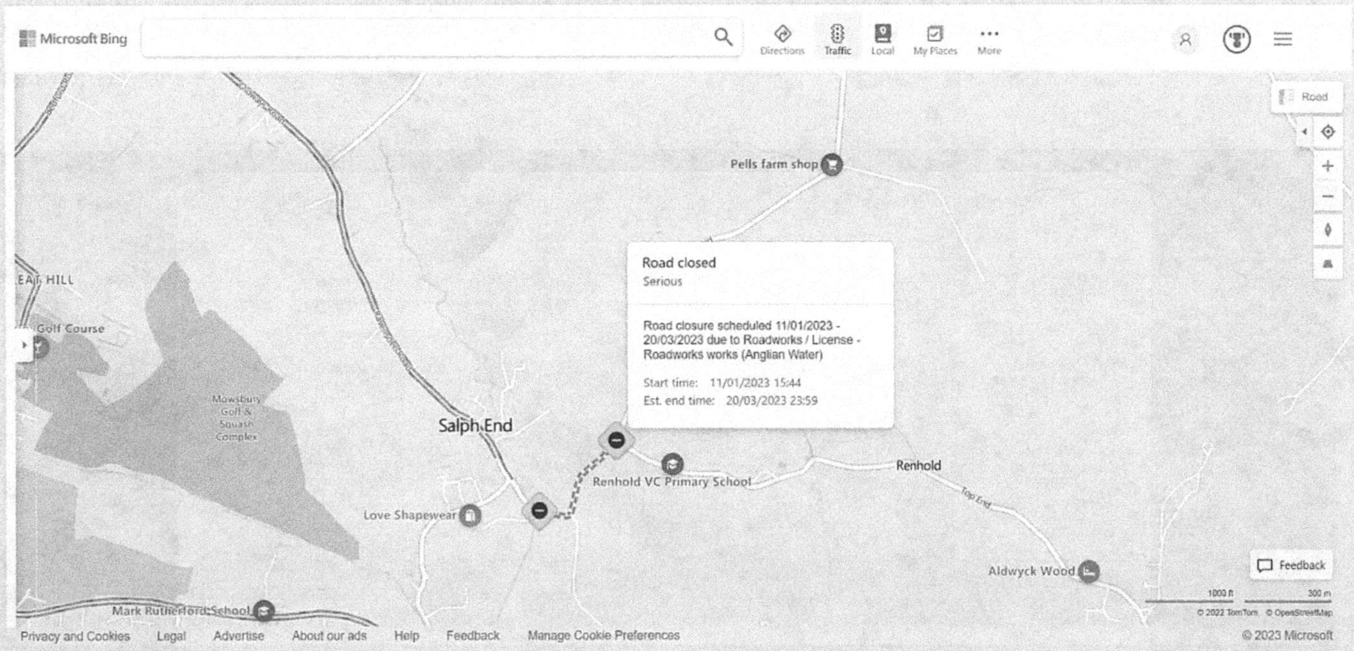

FIGURE 8.7 Functional Test 3: Preconditions (road closure).

This traffic map provided by Bing's web mapping service shows the road closure exploited for the test; the road closure, in combination with the road layout, results in closer vehicles taking longer to reach the emergency location than those further away.

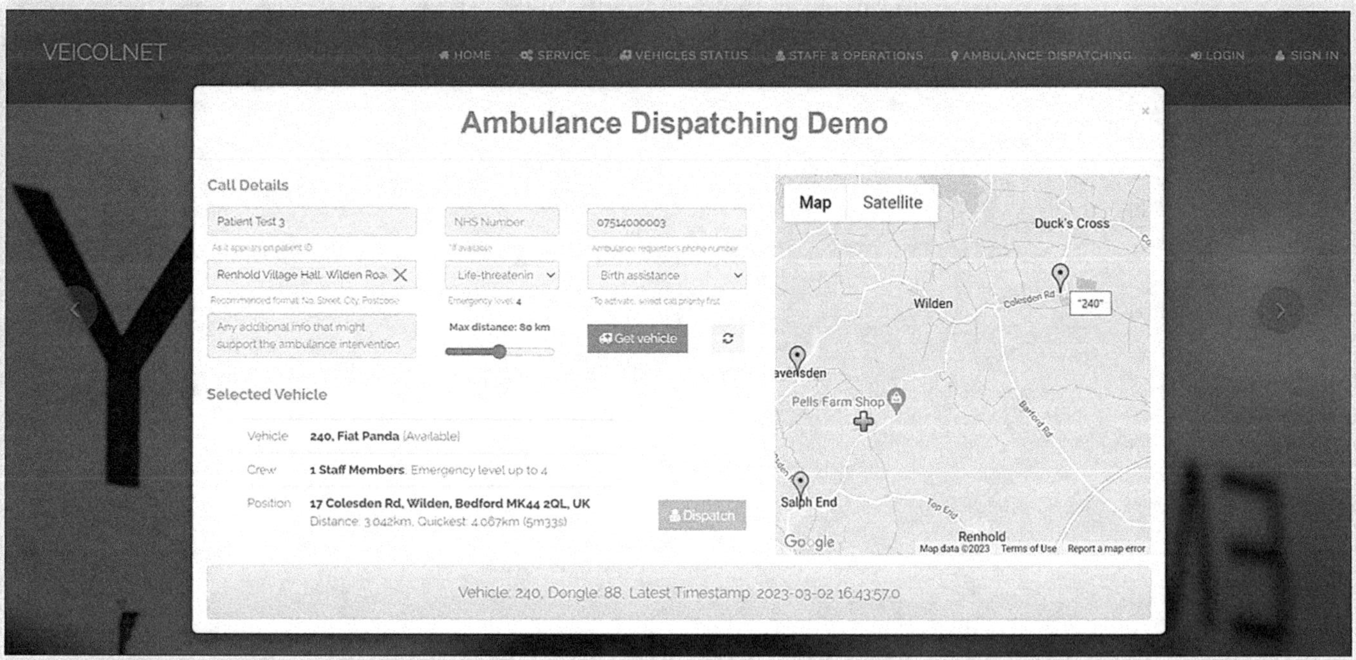

FIGURE 8.8 Functional Test 3: Data input and test result.

This screenshot of the web application UI shows the data used to simulate the emergency and the outcome of the selection process; the functionality of the web app has been modified to display the location of the entire fleet in relation to the destination of the call.

size (number)	3
chosen_vehicle_id (string)	240
list (Array - (3))	

Index	duration	address	distance	level	duration_traffic	driving_distance	vehicle_id	timestamp
0	468	Horse and Jockey, Ravensden, Bedford MK44 2RR, UK	1121.3735272540382	4	474	5562	241	2023-03-02 16:43:57.0
1	676	26 Home Cl, Renhold, Bedford MK41 0LB, UK	1141.9147225822278	4	645	7714	242	2023-03-02 16:43:57.0
2	318	17 Colesden Rd, Wilden, Bedford MK44 2QL, UK	3042.285753016458	4	335	4067	240	2023-03-02 16:43:57.0

FIGURE 8.9 Functional Test 3: Linear distance and driving time.

This figure shows the data the selection is based on; it reveals that the linear distance between the vehicles and the destination, obtained with GIS, is not a reliable indicator, since a shorter distance does not necessarily correspond to a shorter driving time.

As the images (Figures 8.6–8.9) above reveal, our data platform is capable of selecting the vehicle with the shortest driving time to the destination – in this case, vehicle 240. This stands in stark contrast to what would be obtained with the FMS that are currently in use in the control centres of the NHS, which select ambulances based on their linear distance from the emergency, obtained with GIS. As can be seen from the data in Figure 8.9, this approach would have led to the activation of vehicle 241, which not only requires more time to reach the destination than the selected ambulance but would also exceed 7 minutes, surpassing the time target set on a life-threatening call. One might argue that the experience of the call dispatcher – who has the final word in the decision-making process – could have corrected the outcome of the dispatch by selecting the correct vehicle; however, as shown by Figure 8.8, the spatial layout of the fleet provides no indication that vehicle 240 is the fastest to reach the destination, as it is considerably farther away than other vehicles: consequently, the dispatcher would still have been likely to select the wrong ambulance (Table 8.3).

As this test reveals, the CVP not only eliminates both the errors and delays that a human operator introduces into the dispatching procedure, but also allows for better decision-making than the one that is currently possible in the control centres of the NHS. What makes this possible is the aggregation of situation data – namely, vehicle measurements and context data – with live insights from other sources, such as traffic data provided by Google Maps. This information, which is presently unused, is in fact capable of overturning the outcome of the procedure as it is performed today by call dispatchers. Clearly, our choice to rely on a more extensive dataset plays a crucial role in enabling this capability; equally important is our decision to opt for technologies that allow this data to be collected and processed with extreme responsiveness.

Decisive to the result obtained is, for example, our decision to use an OBD dongle that transmits data over MQTT. Adopting an HTTP-based dongle would have required batch communications at relatively low transmission frequencies – on the order of a few tens of seconds, as discussed in Chapter 4. This would have effectively negated any advantage of aggregating traffic data in real-time, as the GPS positions used for routing decisions would have been several seconds old, and therefore not representative of the actual state of the fleet. Just as decisive is the choice of the database, which allows this data to be ingested, processed, and made available for decision-making at high speed.

The ability to combine vehicle measurements, which conventional FMS rely on, context data, used in control centres but not integrated with vehicle data, and live information from other sources, which is currently neglected, would even open up new possibilities in terms of dispatching procedures. For example, rather than selecting the nearest ambulance to the emergency, NHS control centres could activate the vehicle with the lowest skill level among those that are both able to clinically handle the emergency and reach the destination on time. As a result, if there were a vehicle with a paramedic in the immediate vicinity of the destination of an urgent call, it would be ignored in favour of an ambulance with a lower skill level, which, despite being at a greater distance from the emergency, would still be able to reach the call destination in less than 120 minutes. This would allow control centres to retain vehicles with high skill levels for critical emergencies, following a performance model that goes beyond the mere achievement of time targets. The one described is just an example of a decision-making process that is neither feasible by human operators – given its complexity – nor by the existing FMS, incapable of handling its data. Hence the need for CVPs that are tailored to the requirements of the applications to be served.

8.3 VALIDATION OF TEST RESULTS

Within the context of an emergency, *response time* is defined as the interval between the collection of sufficient information about the emergency and the arrival of the first vehicle on the scene. This measure is objective and quantifiable, as well as easy to understand for both decision-makers and the public; as such, it is typically used as a process indicator to assess the performance of the emergency services. Hence, to validate the results of the tests presented in the previous pages, it would be reasonable to explore the ability of the CVP to reduce response times compared to conventional dispatching systems. Our prototype, in fact, guarantees a reduction in the *activation time* of the fleet, as it accelerates vehicle selection, in its *turnout time*, as it avoids delays in the communication between control centres and on-board crews, and in its *travel time*, as it activates the units with the shortest driving distance to the destination.

The performance advantages are illustrated in Figure 8.10, which compares the response times of the systems currently used in the control centres of the San Francisco Police Department - used as a proxy in the absence of an openly available ambulance dataset - with those obtained using our prototype. While the former activate vehicles based on the stations responsible for the area where emergencies occur, the latter selects the vehicle with the shortest driving time to the destination, regardless of its area of responsibility. The ability of our prototype to reduce response times stems from the automation of the decision-making process of control centre operators; this automation, on the one hand, eliminates the delays in activation and turnout times that call dispatchers inevitably introduce and, on the other hand, enables the adoption of more sophisticated dispatching algorithms than those currently in use. Naturally, the actual reduction in travel time would depend on the specific algorithm employed.

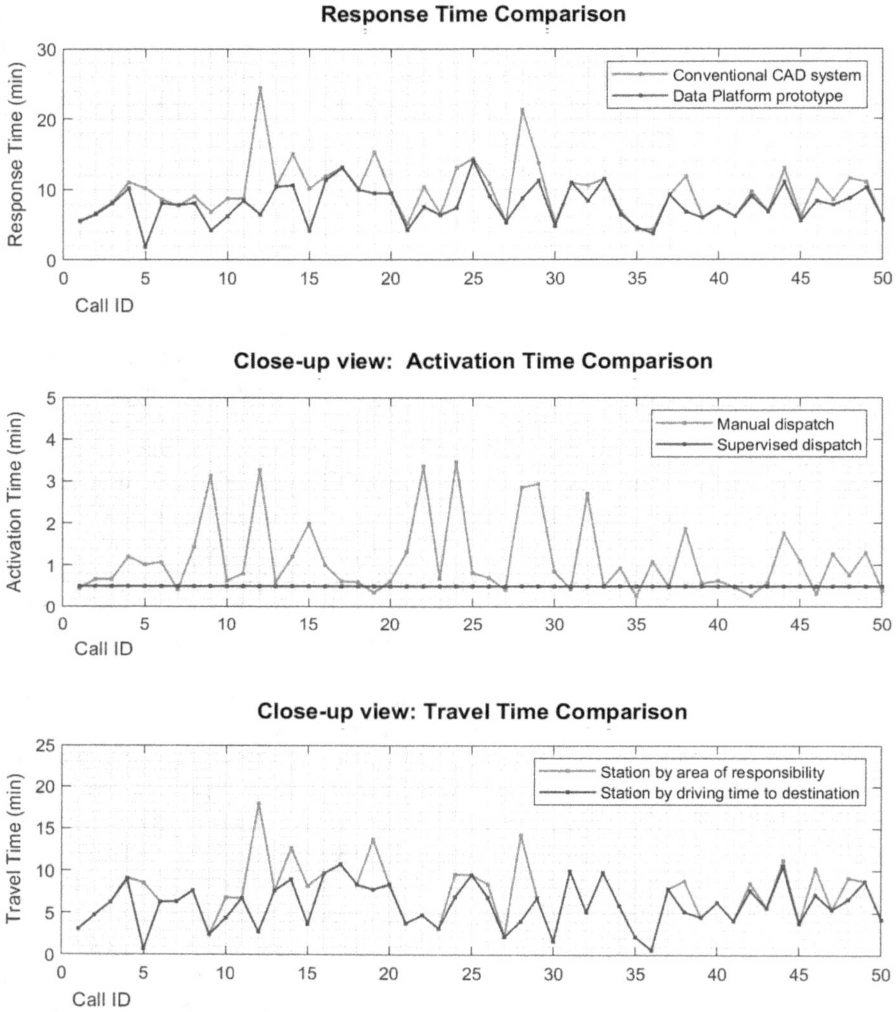

FIGURE 8.10 Response time reduction for the SFPD fleet.

This figure illustrates the improvements in the responsiveness of the emergency fleet that can be achieved by automating the decision-making process of control centre operators. The graphs refer to high-priority calls received from 2 pm to 8 pm on 3 September 2023: these calls have a negligible chance of vehicle unavailability, since they take precedence over ongoing interventions, provided the latter have a lower priority; this ensures that any delays in the activation time are mainly attributable to the tasks of the call dispatchers. The graphs also assume that the call dispatcher spends 30 seconds reviewing the outcome of the automated vehicle selection procedure when using our data platform prototype.

However, confining the validation to a performance comparison based solely on response times would be reductive for this research. The possibility of employing superior algorithms to manage vehicle operations and the elimination of delays in the activities of the human operators are just two of the advantages of automated decision-making. Our CVP, in fact, also mitigates the errors that the operators commit when making decisions in emergency control centres. Consequently, the significance of our test results does not lie in the ability of the prototype to reduce response times, but rather in the ability to do so while limiting the error of the call dispatchers. Indeed, as of today, these two objectives are not only incompatible but even conflicting: any attempts to shorten response times typically lead to higher chances of errors in the dispatching process, posing greater risks for patients and higher costs for NHS trusts.

Having already demonstrated the ability of our prototype to reduce response times, the remainder of this section will focus on its ability to mitigate dispatcher errors. This section opens with a *human factors analysis*, which allows us to identify the factors that contribute to the complexity of decision-making in emergency control centres; this analysis reveals that it is precisely the preoccupation with meeting response time targets, together with the systems in use in control centres, that leads to numerous errors and operational inefficiencies. This is followed by a *heuristic evaluation* based on Gerhardt-Powals' principles of cognitive engineering, which serves as a starting point to identify which of the complexity factors discovered with the human factors analysis can be mitigated, or eliminated altogether, through the use of our prototype. This evaluation concludes our research by demonstrating that all performance advantages offered by our CVP over conventional dispatching systems – from reducing vehicle response times to mitigating errors and delays in the decision-making process – can be attributed to a reduction in the cognitive load of control centre operators.

8.3.1 Human Factors in Control Centre Decision-Making

Rapid response to emergencies is a strong public expectation that NHS trusts, like any other emergency service providers around the world, strive to meet. However, these aspirations often lead trusts to be more concerned with the speed of response than with the quality of care provided to patients. To date, in fact, the performance of the ambulance services is evaluated through time targets whose correlation with better patient outcomes has yet to be established outside of a small subset of calls, such as cardiac and respiratory arrests. While it is acknowledged that these quality indicators can support such organisations in identifying areas of deficiency and ensure both accountability and transparency towards the public and founders, they produce unintended effects; in fact, as emerges from the human factors analysis below, the preoccupation with meeting response time targets, in combination with the systems in place, leads to errors in the decision-making process of control centres, which end up creating systemic dysfunctions in the emergency service.

Unfortunately, the collateral effects of conventional approaches to evaluating emergency service performance remain largely unexplored in research. Emerging studies, however, reveal that most performance indicators in public sectors are designed with the intention of yielding efficiency gains, yet without accounting for the potential costs and repercussions of these schemes. One notable side-effect of the current performance frameworks is *measure fixation*, which is defined as an emphasis on measures of success rather than the underlying objectives; there is, in fact, an alarming tendency among emergency service providers to focus more on performance indicators per se than the intended outcomes. This tendency is evident in the control centres, where trusts frequently choose to activate multiple vehicles for high-priority interventions, maximising the chances of achieving their time targets. This practice, commonly referred to as over-dispatching, not only contributes to increased stress and fatigue for ambulance crews but also leads to additional costs for trusts, like fuel and operational expenses. Moreover, it results in an inefficient allocation of resources, since vehicles are diverted from other areas or incidents where they may be needed more urgently, with a potential shortage of staff for subsequent interventions. Yet, within the current performance framework, this is not regarded as an error but a necessary move, which may, however, compromise service efficiency and patient outcomes. The root cause of this phenomenon lies in the incomplete information available to conventional dispatching systems: despite knowing the location of ambulances, dispatchers do not have an accurate indication of the time it takes for the units to reach their destination.

Alongside this phenomenon is a broad range of accidental errors that are typically attributed to the human component on which the operations of emergency control centres depend. While the pressure to meet response time targets is not the primary cause of these errors, it is certainly a major contributing factor to their occurrence. According to Wiegmann and Shappell, operator performance results from a complex interaction of several factors: (i) the *liveware*, that is the human, (ii) the *system* at their disposal, which includes hardware and software, and (iii) *environmental conditions*. From this perspective, an error occurs when the operator fails to process information appropriately, which may result from a system that does not meet their needs or from environmental and task-related stressors that impair their cognitive capabilities. In control centres, errors attributable to the call dispatcher arise from an interplay of these factors. On one hand, the time-critical environment imposes fast and frugal decisions, a situation that is exacerbated by the number of calls to be handled and the difficulty of expanding the control centre's workforce. On the other hand, current CAD systems place significant demands on operators' cognitive resources, due to the incomplete, inaccurate, and fragmented information they provide. Adding volatility to this scenario is the operator's experience, which further influences the quality of decision-making.

An analysis of the errors that can occur in emergency medical centres is provided by Mortato et al., who reveal that as many as 78% of these are due to human factors, with almost half of them classified as either moderate or severe. The most significant aspect to emerge from this research is, however, the problem of under-reporting, as evidenced by an error rate of only 3.5 errors per thousand operations. Indeed, there is a cultural resistance to event reporting in medicine, primarily stemming from the fear of punitive perception from the public, concerns about damage to professional reputations, and media portrayal of errors. Hence, it is not possible to obtain an exhaustive list of control centre errors, as the events recorded by incident report systems end up capturing a limited view of the whole picture.

Nevertheless, it is still possible to determine which factors contribute to the complexity of pre-hospital emergency decision-making, whose interactions slow down operations and increase the likelihood of errors. The most significant research in this respect is that of J. Hayes et al., which focuses on the operators of two New Zealand control centres differing in their dispatching technology, population served, geographical area covered, and road network layout. This research proves extremely relevant to our human factors analysis, as these control centres – except for minor differences in the technological set-up – are based on an operative framework that is very similar to that in which NHS trusts operate. This study presents 38 factors that, according to the perception of the call dispatcher, contribute to the occurrence of errors in the decision-making process. These factors, listed in the tables below, are ordered by a decreasing Complexity Index (CI), which takes into account not only the complexity but also the frequency of each contributing factor (Tables 8.4 and 8.5).

TABLE 8.4
Factors Contributing to Decision-Making Errors, High CI

CI band	Centre A	Centre B	U	Sig.
CI ≥ 8		High workload (5)	40.5	0.131
		Time pressure (8)	42	0.166
	Area-specific allocation procedures (9)			
	Uncertain location (12)	Uncertain location (12)	56	0.605
	Determining priority (16)*			
	Incident awareness (17)			
		Quickest vehicle (18)	36	0.077
		Quickest additional vehicles (19)*	33.5	0.049
Next available resource (24)	Next available resource (24)	63	0.927	
		Determining crew status (26)	37.5	0.088
		Providing coverage (27)*	2.5	0.000
		Location of resources in transit (28)*	19.5	0.003
		Managing crew workload (29)	56	0.605
		Managing patient transfers (36)	44.5	0.357
7 ≤ CI < 8		Traffic congestion (2)*	23.5	0.017
	High workload (5)			
	Event unpredictability (6)	Event unpredictability (6)	48	0.456
	Rate of change (7)	Rate of change (7)	56.5	0.896
		Incident-specific response procedures (10)	58	0.693
	Uncertain patient condition (13)			
		Information collation (14)	35	0.067
	Not talking with caller (15)	Not talking with caller (15)	55	0.563
		Incident awareness (17)	39	0.115
		Matching crew to incident (20)	50	0.376
	Ensuring the required resources at incident (25)			
		Radio dead spots (30)	46	0.257
		Fatigue (35)	59.5	0.738
	Managing patient transfers (36)			

Note: The values in the tables are used to determine whether there is a significant difference between the centres: while *U* measures the total number of times an observation ranks higher in one of the groups, *sig*. denotes the statistical significance of the Mann-Whitney U-test.

* Denotes significance difference $p < 0.05$.

TABLE 8.5
Factors Contributing to Decision-Making Errors, Low CI

CI Band	Centre A	Centre B	U	Sig.
6 ≤ CI < 7		Number of possible routes (3)	50.5	0.539
	Limited access (4)			
	Time pressure (8)			
		Area-specific allocation procedures (9)	51	0.410
	Incident-specific response procedures (10)			
		Uncertain patient condition (13)	57	0.872
	Information collation (14)			
	Quickest vehicle (18)			
	Quickest additional vehicles (19)*			
	Matching crew to incident (20)			
	Unfamiliar with area (21)	Unfamiliar with area (21)	64.5	0.976
		Low number of dispatch alternatives (23)	46	0.257
	Managing crew workload (29)			
	Radio dead spots (30)			
		Equipment malfunctions (31)	36	0.077
	Fatigue (35)			
	Other distractions (37)	Other distractions (37)	50.5	0.539
CI < 6	Road works/road conditions (1)	Road works/road conditions (1)	55.5	0.563
	Traffic congestion* (2)			
	Number of possible routes (3)			
		Limited access (4)	35.5	0.107
	Low workload (11)*	Low workload (11)*	32.5	0.042
		Determining priority (16)*	32	0.042
	High number of dispatch alternatives (22)	High number of dispatch alternatives (22)	47	0.471
	Low number of dispatch alternatives (23)			
	Determining crew status (26)			
	Providing coverage (27)*			
	Location of resources in transit (28)*			
	Equipment malfunctions (31)			
	Equipment distraction (32)	Equipment distraction (32)	42.5	0.166
	Joint responses with other regions (33)	Joint responses with other regions (33)	34	0.210
	Different terminology across services (34)	Different terminology across services (34)	39	0.115
	Training exercises (38)	Training exercises (38)	60	0.784

Note: The values in the tables are used to determine whether there is a significant difference between the centres: while U measures the total number of times an observation ranks higher in one of the groups, *sig*. denotes the statistical significance of the Mann-Whitney U-test.

* Denotes a significant difference, $p < 0.05$.

The results of this study do nothing but confirm the hypotheses formulated in the previous pages. Firstly, the quality of decision-making is highly dependent on the environmental conditions in which dispatchers operate: time pressure and high workload are determining factors for the occurrence of errors, with an impact that varies with the volume of calls to be handled. Secondly, the decision-making process requires significant cognitive effort, as the current dispatching systems are not able to provide operators with a clear mental picture of the emergencies to be managed; dispatchers, in fact, are often required to collate, compare, and mentally integrate information from various sources, which is frequently incomplete, outdated, or presented out of context. Even the use of well-known and accepted heuristics, such as activating the nearest ambulance or sending an advanced crew to a high-priority emergency, does not avoid decision-making errors, as these strategies are only effective in simple situations. In more complex, but fairly common scenarios, such as when the closest ambulance capable of providing advanced assistance is further away than other vehicles, it is not always so easy to determine the best of the available sub-optimal options. In addition, recourse to these mental shortcuts as part of a subjective, anecdotal, and experience-based decision-making can result in cognitive biases, which may lead to erroneous choices.

The quality of decision-making in emergency control centres thus depends on the cognitive load of the operators, which is affected by the conditions they operate in and the systems they use. As the former can hardly be influenced – with workload and time pressure even set to worsen – the only solution to reduce the occurrence of errors is to enhance the human-machine interface. Yet, it is not simply a matter of using more powerful machines, as this would not bring significant benefits. The key is to use a cognitively ergonomic system, which is able to gather, collate, and present all the data required for decision-making in a timely manner.

8.3.2 Cognitive Ergonomics and Service Performance

As evidenced by the human factors analysis presented in the previous section, the errors that occur in emergency control centres can be traced back to a lack of understanding of the cognitive needs of the operator, as well as the inability to respond to these needs through the human-machine interface. Addressing this issue is imperative, considering that over 90% of these errors – which have a considerable impact on the performance of emergency services – are entirely preventable. The data platform developed as part of this project marks an important step in this direction, as it mitigates, and in some cases eliminates altogether, the factors that can negatively affect the quality of decision-making. Achieving this, however, is not simply a matter of screen formats and data presentation; it is the result of the reasoned design of the whole CVP, whose capabilities surface in its user interface.

To prove the effectiveness of our prototype in reducing call dispatcher errors, we opted for an analysis based on the *Gerhardt-Powals principles*, which draw on concepts such as *cognitive ergonomics* and *situation awareness*. Unlike the more commonly used Nielsen heuristics, – which primarily help identify flaws and usability issues in the user interface – these principles allow us to assess the impact of the entire data platform on the cognitive load of the operators. Table 8.6 outlines how the CPV aligns with Gerhardt-Powals principles in comparison to conventional CAD systems. This analysis serves as a starting point for determining which factors contributing to the occurrence of errors can be mitigated through the use of our CVP. To quantify error reduction, a usability test would be required: this would require deploying our prototype in an actual control centre, so as to allow a direct comparison with existing systems. However, this test entails a commitment in terms of cost, time, and user involvement that is beyond the scope of this project; therefore, it is deferred to future research.

TABLE 8.6
Cognitive Ergonomics Evaluation of the CVP Through Heuristics

Gerhardt-Powals Principles	CVP Features for Vehicle Dispatch
1. Automate unwanted workload: Eliminate mental calculations, comparisons, estimations, and any other superfluous thinking, so as to free cognitive resources for high-level tasks	• Automatic vehicle selection • Automatic call status management • Automatic staff status management • Automatic communication with vehicle staff
2. Reduce uncertainty: Display data in a manner that is clear and obvious so as to reduce decision time and errors	• Vehicle selection procedure displaying only the unit with the shortest travel time to the destination that meets all dispatching criteria • Data on vehicle travel time to destination
3. Fuse data: Combine low-level data into high-level summaries so as to reduce the cognitive effort of the user	Unified dashboard providing vehicle data (geospatial and driving data), emergency call details, and personnel status information.
4. Simplify presentation of new data: So that information is easier to absorb	Context data on calls and staff is presented on a map together with vehicle data.
5. Use names according to function: So as to improve recall and recognition	Context data on calls and staff is presented, retaining both the format and terminology of conventional dispatching systems.
6. Group data: So as to reduce information search time	Data is organised by entity, with dedicated screen sections for vehicles, calls, and staff.
7. Limit data-driven tasks: So as to reduce the time to assimilate raw data	Vehicle location is displayed on a map with markers of different colours depending on the status of the current call and that of its staff.
8. Show only the information needed: So that the user can focus attention on critical data	Although the CVP captures all data useful for fleet management, only the data that is relevant for vehicle dispatching is shown to the user.
9. Provide multiple coding of data: So as to promote cognitive flexibility	Fleet status monitoring via two different view modes (multi-vehicle, single-vehicle).
10. Practice judicious redundancy: To resolve possible conflicts between 6 and 8	The data represented on the map is simultaneously given in text form.

Note: The table outlines the unique ways our prototype aligns with Gerhardt-Powals principles. This analysis allows us to determine the way our CVP reduces the cognitive load of the operators compared to the dispatching systems currently used in control centres.

As the heuristic evaluation reveals, the edge of our data platform over conventional dispatching systems lies in the automation of a substantial portion of the workload of control centre operators. By automating tasks that are currently performed manually, the prototype relieves the operators of the cognitive demands associated with complex decision-making processes and other routine actions. Most notably, our data platform relieves call dispatchers from the responsibility of selecting the ambulance to be activated; this is achieved through the unique ability of our CVP to gather all the data required for decision-making, enabling existing vehicle selection procedures to be executed without human intervention. Automating this task removes many of the factors that presently contribute to errors and delays in the decision-making process of emergency control centres; these factors are listed below, along with the numbers by which they appear in the tables of Section 8.3.1. For instance, the ability of the CVP to jointly process vehicle data and context data on staff and calls, which is not possible with current CAD systems, overcomes the challenge that dispatchers face in matching ambulance crews to emergencies (20). The processing of context data that is currently unused, such as driving time data, eliminates difficulties in locating the quickest ambulance (18) and the next available resources (24), determining the best route to the destination (3), and assessing the presence of road works (1) or traffic congestion (2) along the route. Our data platform even allows for different dispatching procedures to be automatically applied depending on the area (9) and the type of emergency (10), removing two of the most significant sources of complexity in control decision-making. This automation also prevents over-dispatching, limits the influence of operator experience on the quality of decisions, and avoids delays in vehicle mobilisation.

Access to a unified dataset that includes both vehicle and context data, coupled with the ability to delegate its processing to an automated decision-maker, even allows for more sophisticated dispatching algorithms than those currently in use. Indeed, many of the vehicle selection strategies proposed in the academic literature cannot be adopted in today's control centres due to the poor cognitive ergonomics of existing CAD systems. A significant portion of the cognitive load would fall on call dispatchers, who would be required to perform complex calculations, comparisons, and estimates – tasks that are infeasible given the pressing need to meet response time targets. Implementing such algorithms would yield substantial benefits, as they not only reduce ambulance response times but also improve fleet performance in numerous other ways, as discussed in Section 2.2.4.

As shown in Table 8.6, our prototype also automates the management of call status and ambulance staff status. These activities are a clear example of auxiliary tasks that, while not directly related to decision-making, still contribute to the cognitive load of control centre operators. Similar to conventional CAD systems, the creation of a call record in the database is automatically performed by the call dispatcher's user interface following call triage. However, call termination is no longer the responsibility of the control centre; it is delegated to the ambulance operators on board the vehicle that responded to the emergency. Likewise, the management of the status of the on-board crew is fully entrusted to ambulance operators through the *shift management function* of their mobile application. Obviously, automating these tasks reduces the likelihood of errors in the dataset that support decision-making, which might otherwise result from operator distraction (37). It also eliminates the challenges faced by dispatchers when it comes to determining the status of the ambulance crews (26) and managing their workload (29). Our data platform even simplifies the exchange of emergency information between the control centre and the ambulance assigned to the call, through automatic notifications between the call dispatcher's web app and the ambulance staff's mobile app. This is an explicit attempt to address the complex socio-technical challenges of the human-human interface, which is also recognised as a major cause of errors in control centres. Naturally, automating these auxiliary tasks also eliminates any delays caused by communication and coordination between operators.

In light of the above, a dispatching system built according to the principles of cognitive ergonomics can even reduce the impact of environmental conditions on the performance of control centre operators, as it lightens their workload (5) and consequently relieves the time pressure (8) they experience. This is a remarkable achievement considering the situation emergency services currently face, as it allows the performance of their control centres to be improved without increasing their workforce – a solution that is not feasible given the limited financial resources available. Obviously, automating the workload transforms the role of dispatchers from decision-makers to supervisors; in this context, it is imperative that the user interface can effectively assist operators in their new responsibilities. For example, dispatchers should be able to easily assess the choices made by the automated decision-maker. To facilitate this, our user interface displays the location of the selected vehicle on a map while simultaneously providing a textual overview of the parameters considered in the decision-making process, including the type of emergency and the skills of the crew involved. It also provides an estimate of the driving time required for the ambulance to reach the destination, which the dispatcher can use to determine whether the unit can meet the response time target, as well as the vehicle's estimated time of arrival, which can be shared with the caller.

In terms of visual appearance, our web application aligns with the user interfaces used by traditional CAD systems. However, its operation differs substantially: it displays both vehicle data, typical of control centre systems, and context data, which is currently fragmented across other systems or even unused. Our prototype, therefore, addresses another challenge faced by call dispatchers, namely that of information collation (14), which requires operators to gather and mentally combine data from multiple systems. In formal terms, our CVP increases the *situation awareness* of call dispatchers, which is achieved through a single interface that provides the right information, in the right format, at the right time. Indeed, as confirmed by the literature, partial solutions add enormous complexity, leading to poor-quality decisions, errors, and delays.

All the performance advantages that our prototype offers over conventional CAD systems can then be attributed to a reduction in the cognitive load of call dispatchers. This improvement in cognitive ergonomics stems from the ability of our data platform to gather, elaborate, and present all the data involved in decision-making with extreme responsiveness. By contrast, the systems in use in control centres – built on a data platform originally intended for traditional fleet management – lack these capabilities. This strongly reinforces the premise that we opened this book with: emergency services require a bespoke CVP, as the management of these vehicles entails non-functional requirements that differ significantly from those of other fleet management applications.

8.4 CHAPTER SUMMARY

This chapter has proven the ability of the CVP to automate the decision-making process behind ambulance dispatching, enabling performance and functionalities that are unattainable with existing emergency FMS. This confirms the hypothesis advanced in this book: context-aware, real-time fleet management requires a tailor-made data platform, as its non-functional requirements significantly deviate from those of other applications.

9 Conclusions

The aim of this research was to implement a CVP capable of supporting context-aware, real-time fleet management applications, which have different data collection and processing requirements compared to typical fleet management. Emergency fleets, taken as a case in point, exemplify this need. By designing the data platform around these requirements, this research led to the development of a prototype for managing the ambulance fleet of the NHS, enabling the automation of the decision-making processes behind vehicle operations and providing performance and functionalities that cannot be achieved with existing systems. This outcome demonstrated the critical influence of the data platform on fleet management, confirming the central hypothesis of this research: each fleet requires its own CVP, tailored to the non-functional requirements of the applications it will serve.

The results of this project revealed the importance of incorporating data platform considerations into fleet management studies. Academic studies, which mainly focus on developing innovative algorithms for fleet management, often overlook the technological characteristics of the CVP on which these algorithms would be deployed. This is a perilous approximation, as it not only prevents a proper evaluation of algorithm performance – given the significant influence of the underlying data platform – but also hampers the practical implementation of these solutions, since no commercial platforms may be capable of properly handling their data. We hope that the findings of this book will help bridge the gap between the academic and commercial worlds in fleet management, so that the pioneering ideas conceived by the research community can become an integral part of our everyday lives.

FUTURE DIRECTIONS

The research undertaken for this book has brought to light a number of topics for which further investigation would be beneficial; our recommendations in this regard are listed below. In addition, we note that this book could also serve as a model for designing data platforms for fleets with different requirements, as it provides a detailed description of the software engineering methodology underpinning the design process.

CVP Performance Testing

Future projects might evaluate the behaviour of the CVP under realistic operating conditions, which would require further development of the data platform prototype. The prototype described in this book, in fact, does not support performance testing, as it was designed solely to meet the requirements necessary for correctly handling a

DOI: 10.1201/9781003688549-9

context-aware, real-time fleet management application. A full-scale, commercial version of the system would need to address additional requirements, such as security, reliability, and scalability, which go beyond the scope of this research.

CVP Usability Testing

Future research might evaluate the performance of the CVP within a real control centre. Conducting a usability test in an operational environment would, in fact, enable a direct comparison with existing solutions and make it possible to quantify the improvements in the quality of decision-making by the emergency response team.

Predictive Maintenance Through Event-Streaming Platforms

Future projects might explore the possibility of using an event-streaming platform to enable predictive maintenance capabilities in the CVP. This research, in fact, merely evaluated the role of these technologies in the implementation of a vehicle malfunction notification service, which supports the decision-making process associated with emergency fleet operations.

Analysis of the Impact of Other Context Data on Ambulance Dispatching

Future studies might assess the impact of additional types of context data on ambulance dispatching, such as air quality monitoring, weather forecasts, and mobile phone location data. This research, in fact, does not consider the full range of context data that could improve decision-making in emergency vehicle operations.

Analysis of the Impact of Context Data on Performance Evaluation Activities

Future research might investigate the effect of combining vehicle data and context data on performance evaluation activities typical of traditional fleet management. This book, in fact, while exploring the implications of improved data management for ambulance dispatching, does not examine the potential benefits that such an approach could bring to other fleet management activities.

Appendices

Appendix A
Database Code

The purpose of Appendix A is to provide an in-depth description of the database outlined in Chapter 4, as the latter merely offers an overview of its characteristics, without delving into the details of its structure. To fill these gaps, this appendix lists the code for creating the database schema, which encodes information on the structure of the individual tables of the database, including the data types of their columns and the constraints they are subject to. These tables, as can be noted from Listing A.1, are created in a different order from the one used to describe them in Chapter 4, as this is required to satisfy the referential integrity constraints of the database. For the sake of completeness, this listing also includes the SQL commands for adding the *TimescaleDB*, *PostGIS*, and *pgcrypto* modules to the database, as well as the code needed to create the *hypertable*. With regard to the latter, the code listing reveals that the only form of partitioning used in this project is *time partitioning*. This choice stems from the fact that the database is hosted on a single node, making the use of spatial partitioning entirely superfluous, at least in this prototype version of the data platform; such a strategy would prove beneficial only when the database is distributed across several physical disks, a situation that could occur in the commercial implementation of the CVP. Following this is Listing A.2, which presents the code used to populate the static tables of the database, namely, those whose records are defined during the configuration phase of the data platform and are not expected to change subsequently. The tables in question are the *Staff_Role* and *Call_Type*, whose records have been generated based on the information derived from official NHS reports on ambulance service management, which are readily available online.

LISTING A.1 SQL CODE FOR CREATING THE DATABASE SCHEMA

```
-- CREATE DATABASE EXTENSION

-- PostGIS --
CREATE EXTENSION postgis;

-- TimescaleDB --
CREATE EXTENSION IF NOT EXISTS timescaledb CASCADE;

-- Pgcrypto --
CREATE EXTENSION pgcrypto;
```

```sql
-- CREATE DATABASE SCHEMA

-- DONGLE --

-- Dongle ID sequence
CREATE SEQUENCE dongle_id_sequence;

-- Dongle table
CREATE TABLE Dongle (
       dongle_id INTEGER DEFAULT NEXTVAL('dongle_id_sequence') PRIMARY KEY,
       mac MACADDR UNIQUE NOT NULL,
       status INTEGER NOT NULL,
       buffer_size INTEGER NOT NULL
);



-- VEHICLE --

-- Vehicle ID sequence
CREATE SEQUENCE vehicle_id_sequence;

-- Vehicle table
CREATE TABLE vehicle (
       vehicle_id INTEGER DEFAULT NEXTVAL ('vehicle_id_sequence') PRIMARY KEY,
       VIN CHAR(17) UNIQUE NOT NULL,
       make VARCHAR(20),
       model VARCHAR(20),
       year_made INTEGER NOT NULL,
       year_registered INTEGER NOT NULL,
       weight INTEGER NOT NULL,
       engine_full VARCHAR(50),
       engine_displ INTEGER NOT NULL,
       fuel_type VARCHAR(10),
       model_id INTEGER NOT NULL,
       CONSTRAINT vehicle_model_id_fkey FOREIGN KEY (model_id)
          REFERENCES emission_model (model_id) MATCH SIMPLE ON
          UPDATE NO ACTION ON DELETE NO ACTION
);


-- EMISSION MODEL --

-- Emission Model ID sequence
CREATE SEQUENCE model_id_sequence;
```

```
-- Emission Model table
CREATE TABLE Emission_model (
        model_id INTEGER DEFAULT NEXTVAL ('model_id_sequence') PRIMARY KEY,
        weight_class INTEGER NOT NULL,
        engine_class INTEGER NOT NULL,
        fuel_class INTEGER NOT NULL,
        emission_class INTEGER NOT NULL,
        CO_values FLOAT[8],
        HC_values FLOAT[8],
        NOx_values FLOAT[8],
        PM_values FLOAT[8]
);


-- MEASUREMENT --

-- Measurement ID sequence
CREATE SEQUENCE measurement_id_sequence;

-- Measurement table
CREATE TABLE Measurement (
        measurement_timestamp TIMESTAMPTZ NOT NULL,
        dongle_id INTEGER NOT NULL,
        PRIMARY KEY(measurement_timestamp, dongle_id),
        vehicle_id INTEGER NOT NULL,
        lat FLOAT NOT NULL,
        lon FLOAT NOT NULL,
        GPS_position GEOMETRY(Point, 4326),
        speed INTEGER NOT NULL,
        engine_rpm INTEGER NOT NULL,
        CO FLOAT NOT NULL,
        HC FLOAT NOT NULL,
        NOx FLOAT NOT NULL,
        PM FLOAT NOT NULL,
        CONSTRAINT measurement_dongle_id_fkey FOREIGN KEY (dongle_id)
        REFERENCES dongle (dongle_id) MATCH SIMPLE
            ON UPDATE NO ACTION ON DELETE NO ACTION,
        CONSTRAINT measurement_vehicle_id_fkey FOREIGN KEY (vehicle_id)
             REFERENCES vehicle (vehicle_id) MATCH SIMPLE
             ON UPDATE NO ACTION ON DELETE NO ACTION
);

-- STAFF ROLE --

-- Staff role table
CREATE TABLE staff_role (
        vehicle_role_id INTEGER PRIMARY KEY,
        title VARCHAR(30) UNIQUE NOT NULL,
        acronym VARCHAR(10) UNIQUE NOT NULL
);
```

```
-- STAFF --

-- Staff ID sequence
CREATE SEQUENCE staff_id_sequence;

-- Staff table
CREATE TABLE staff (
        staff_id INTEGER DEFAULT NEXTVAL ('staff_id_sequence') PRIMARY KEY,
        vehicle_role_id INTEGER NOT NULL,
        staff_number CHAR(7) UNIQUE NOT NULL,
        staff_password TEXT UNIQUE NOT NULL,
        staff_name VARCHAR(20) NOT NULL,
        staff_surname VARCHAR(20) NOT NULL,
        staff_phone VARCHAR(20) NOT NULL,
        CONSTRAINT staff_vehicle_role_id_fkey FOREIGN KEY (staff_role_id)
             REFERENCES staff (vehicle_role_id) MATCH SIMPLE
             ON UPDATE NO ACTION ON DELETE NO ACTION,
);


-- SHIFT --

-- Shift ID sequence
CREATE SEQUENCE shift_id_sequence;

-- Shift table
CREATE TABLE shift (
        shift_id INTEGER DEFAULT NEXTVAL ('shift_id_sequence') PRIMARY KEY,
        start_timestamp TIMESTAMPTZ NOT NULL,
        end_timestamp TIMESTAMPTZ,
        staff_id INTEGER NOT NULL,
        vehicle_id INTEGER NOT NULL,
        CONSTRAINT shift_staff_id_fkey FOREIGN KEY (staff_id)
            REFERENCES staff (staff_id) MATCH SIMPLE
            ON UPDATE NO ACTION ON DELETE NO ACTION,
        CONSTRAINT shift_vehicle_id_fkey FOREIGN KEY (vehicle_id)
            REFERENCES vehicle (vehicle_id) MATCH SIMPLE
            ON UPDATE NO ACTION ON DELETE NO ACTION
);


-- CALL TYPE --

-- Call type ID sequence
CREATE SEQUENCE calltype_id_sequence;

-- Call type table
CREATE TABLE calltype (
```

```
        calltype_id INTEGER DEFAULT NEXTVAL ('calltype_id_sequence') PRIMARY
        ↪ KEY,
        priority INTEGER NOT NULL,
        details VARCHAR(40) NOT NULL
);


-- CALL --

-- Call ID sequence
CREATE SEQUENCE call_id_sequence;

-- Call table
CREATE TABLE call (
        call_id INTEGER DEFAULT NEXTVAL ('call_id_sequence') PRIMARY KEY,
        calltype_id INTEGER NOT NULL,
        vehicle_id INTEGER,
        start_timestamp TIMESTAMPTZ NOT NULL,
        end_timestamp TIMESTAMPTZ,
        service VARCHAR(60),
        address VARCHAR(100) NOT NULL,
        lat FLOAT NOT NULL,
        lon FLOAT NOT NULL,
        GPS_position GEOMETRY(Point, 4326),
        patient_nhsnumber VARCHAR(10),
        patient_fullname VARCHAR(40) NOT NULL,
        patient_phone VARCHAR(20) NOT NULL,
        additional_info TEXT;
        CONSTRAINT call_calltype_id_fkey FOREIGN KEY (calltype_id)
            REFERENCES calltype (calltype_id) MATCH SIMPLE
            ON UPDATE NO ACTION ON DELETE NO ACTION,
        CONSTRAINT call_vehicle_id_fkey FOREIGN KEY (vehicle_id)
             REFERENCES vehicle (vehicle_id) MATCH SIMPLE
             ON UPDATE NO ACTION ON DELETE NO ACTION
);

-- CREATE HYPERTABLE (Temporal Data Chunking)
SELECT create_hypertable('measurement', 'measurement_timestamp');
```

LISTING A.2 SQL CODE FOR POPULATING STATIC DATABASE TABLES

```
-- POPULATE STAFF ROLE: ID is used as Competence Level
-- 1: Patient Transport Service (PTS)
-- 2: Urgent Care Assistant (UCA)
-- 3: Emergency Medical Technician (EMT)
-- 4: Paramedic (Paramedic)

INSERT INTO staff_role (staff_role_id, title, acronym)
VALUES ('1', 'Patient_Transport_Service', 'PTS');

INSERT INTO staff_role (staff_role_id, title, acronym)
VALUES ('2', 'Urgent_Care_Assistant', 'UCA');

INSERT INTO staff_role (staff_role_id, title, acronym)
VALUES ('3', 'Emergency_Medical_Technician', 'EMT');

INSERT INTO staff_role (staff_role_id, title, acronym)
VALUES ('4', 'Paramedic', 'Paramedic');
-- POPULATE CALL: below are the four possible priority levels for an emergency
-- 1: non-urgent
-- 2: urgent
-- 3: emergency
-- 4: life-threatening

-- NON URGENT CALLs
INSERT INTO calltype (priority, details)
VALUES ('1', 'Diarrohea');

INSERT INTO calltype (priority, details)
VALUES ('1', 'Vomiting');

INSERT INTO calltype (priority, details)
VALUES ('1', 'Non-traumating back pain');

INSERT INTO calltype (priority, details)
VALUES ('1', 'Hospital Transfer');

INSERT INTO calltype (priority, details)
VALUES ('1', 'Other non-urgent condition');


-- URGENT CALLS
INSERT INTO calltype (priority, details)
VALUES ('2', 'Fainting (now alert)');

INSERT INTO calltype (priority, details)
VALUES ('2', 'Diabetic problem');
```

```
INSERT INTO calltype (priority, details)
VALUES ('2', 'Isolated limb fracture');

INSERT INTO calltype (priority, details)
VALUES ('2', 'Abdominal pain');

INSERT INTO calltype (priority, details)
VALUES ('2', 'Other urgent condition');


-- EMERGENCY CALLS
INSERT INTO calltype (priority, details)
VALUES ('3', 'Fainting (not alert)');

INSERT INTO calltype (priority, details)
VALUES ('3', 'Acute myocardial infarction');

INSERT INTO calltype (priority, details)
VALUES ('3', 'Major burn');

INSERT INTO calltype (priority, details)
VALUES ('3', 'Sepsis');

INSERT INTO calltype (priority, details)
VALUES ('3', 'Other emergency condition');

-- LIFE-THREATENING CALLS
INSERT INTO calltype (priority, details)
VALUES ('4', 'Cardiac arrest');

INSERT INTO calltype (priority, details)
VALUES ('4', 'Birth assistance');

INSERT INTO calltype (priority, details)
VALUES ('4', 'Acute severe asthma');

INSERT INTO calltype (priority, details)
VALUES ('4', 'Allergic reaction');

INSERT INTO calltype (priority, details)
VALUES ('4', 'Other life-threatening condition');



-- POPULATE EMISSION MODEL:
-- The parameters of the emission models are automatically extracted
-- from the spreadsheets available on the NAEI website [90].
```

Appendix B

Communication Infrastructure Code

The purpose of Appendix B is to outline the functioning of the communication infrastructure of the CVP, already introduced in Chapter 5. Presented below is Listing B.1, which describes the functions used by the API to enable the exchange of context data via HTTP between the various components of the data platform and the database. Following this is Listing B.2, which illustrates the mechanism through which the HiveMQ broker stores the vehicle data obtained over MQTT in the storage unit. The decision to represent the JavaScript software by means of pseudocode derives from the realisation that both scripts – also due to the SQL queries they incorporate – would otherwise be extremely difficult to read.

LISTING B.1 HTTP API FUNCTIONS FOR CONTEXT DATA TRANSMISSION

```

//API 1: Dongle Configuration
//Info: It returns the IDs of the dongle and the vehicle to be configured
//Input:Dongle MAC Address, Vehicle VIN
//Output: Dongle ID, Vehicle ID, Emission Parameters
const getDongleSetupInfo = async (request, response) => {

     //Get input parameters from request message
     GET MAC Address, Vehicle VIN from request

     //Dongle set-up
     VERIFY if the dongle already exists in the DB using MAC address
     if the dongle exists, RETURN its Dongle ID
     else SAVE Dongle in DB, RETURN its Dongle ID

     //Vehicle set-up
     VERIFY if the vehicle already exists in the DB using Vehicle VIN
     if Vehicle exists, RETURN its Vehicle ID
     else
          //Decode VIN using an external API
          GET Vehicle Info using VIN Decoder API
          if some information is missing, EXTRACT it locally from the VIN

          //Choose the emission model to be assigned to the vehicle
          DETERMINE emission model ID based on vehicle info via local processing

```

```
            //Save the vehicle in the DB and associate it with its emission model
            SAVE Vehicle in DB, RETURN Vehicle ID and Emission Parameters

        //Return operation outcome via response message
        SEND Dongle ID, Vehicle ID, Emission Parameters via response

    //API 2: Ambulance Operator Role Retrieval
    //Info: It returns the list of possible ambulance operator roles
    //Input: -
    //Output: List of ambulance operator titles
    const getStaffRoles = (request, response) => {

        //The resulting list is saved in a JSON array
        GET all ambulance operator role titles stored in the StaffRole table

        //Return operation outcome via response message
        SEND List via response
    }


    //API 3: Ambulance Operator Sign-in
    //Info: It creates a new ambulance operator profile
    //Input: StaffNumber, Password, Name, Surname, Phone, RoleTitle
    //Output: Staff ID
    const saveStaff = async (request, response) => {

        //Get input parameters from request message
        GET StaffNumber, Password, Name, Surname, Phone, RoleTitle from request

        //Obtain ID of the role to be associated with the operator
        GET RoleID from DB via RoleTitle

        //Encryption is achieved by hashing and salting through the blowfish cipher
        ENCRYPT Password

        //Save the profile in the DB and associate it with its staff role
        SAVE Staff in DB, return its Staff ID

        //Return operation outcome via response message
        if StaffNumber already exists in DB, SEND 0 via response
        else SEND Staff ID via response
    }


    //API 4: Ambulance Operator Log-in
    //Info: It retrieves the ambulance operator details from the DB
    //Input: StaffNumber, Password
    //Output: Staff ID, StaffNumber, Name, Surname, Phone, RoleAcronym
    const getStaffData = (request, response) => {

```

```
    //Get input parameters from request message
    GET StaffNumber, Password from request

    //Staff and StaffRole tables are INNER JOINED on Staff ID
    GET Staff ID, StaffNumber, Name, Surname, Phone, RoleAcronym from DB via
    ↪ StaffNumber and Encrypted Password

    //Return operation outcome via response message
    SEND Staff ID, StaffNumber, Name, Surname, Phone, RoleAcronym via response
}

//API 5: Start Ambulance Shift
//Info: It creates a vehicle shift for an ambulance operator
//Input: Staff ID, Vehicle ID, Start Shift Timestamp
//Output: Shift ID,
//Vehicle VIN, Vehicle Make, Vehicle Model,
//Call ID, Call Address, Call Details, Call Priority,
//Patient Full Name, Patient Phone, Call Additional Info
const saveShift = async (request, response) => {

    //Get input parameters from request message
    GET Staff ID, Vehicle ID, Start Shift Timestamp from request

    //Save the shift in the DB and associate it with its operator and vehicle
    SAVE Shift with Start Shift Timestamp, return its Shift ID

    //Vehicle and Call tables are INNER JOINED on Vehicle ID
    //Call is returned if the vehicle has a call with end timestamp null
    GET Vehicle Info and Call info from DB via Vehicle ID

    //Return operation outcome via response message
    SEND Staff ID, Vehicle Info, Call Info via response
}



//API 6: Stop Ambulance Shift
//Info: It stops an open vehicle shift by setting its end-timestamp
//Input: Shift ID, End Shift Timestamp
//Output: Shift Status
const closeShift = async (request, response) => {

    //Get input parameters from request message
    GET Shift ID, End Shift Timestamp from request

    //The query is performed on the record whose primary key is Shift ID
    SET End Timestamp

    //Return operation outcome via response message
    SEND status of operation via response
}
```

```
//API 7: Close Emergency Call
//Info: It closes an open call by setting its service and end-timestamp
//Input: Call ID, Service, End Call Timestamp
//Output: Call Status
const closeCall = async (request, response) => {

     //Get input parameters from request message
     GET Call ID, Service, End Call Timestamp from request

     //The query is performed on the record whose primary key is Call ID
     SET End Timestamp and Service

     //Return operation outcome via response message
     SEND status of operation via response
}
```

LISTING B.2 MQTT PLUGIN FOR VEHICLE DATA PERSISTENCE

```
//MQTT Persistance Plugin:
//Info: Allows the MQTT broker to store a vehicle measurement in the DB
//Input: Measurement
async function saveMeasurement (measurement) {

     //Get input parameters from the received message
     GET Dongle ID, Vehicle ID, Timestamp, Lat, Lon,
     Speed, EngineRPM, CO, HC, NOX, PM from measurement

     //GPS position is calculated via PostGIS using SRID 4236, as per WGS84
     CALCULATE GPS position from lat and lon

     //The measurement is not stored if a record with the same primary key
     ↪ exists in the DB; the conflict can arise since a QoS of 1 is used
     if the measurement already exists in DB, DO nothing
     else SAVE Measurement in DB using input parameters and GPS position
}
```

Appendix C
Vehicle Interface Code

The purpose of Appendix C is to provide an overview of the approach we adopted to implement the software of the dongle, whose functionalities have already been discussed in Chapter 6. Presented below is Listing C.1, which illustrates the way SocketCAN has been used to manage the transmission and reception of ISO 15765 messages on CAN. This is followed by Listing C.2, which contains the algorithm to request the VIN from the vehicle's diagnostic control unit; the latter is presented in the form of pseudocode, since the source code, especially if considered in isolation, would have been particularly difficult to understand. Listing C.3 features, instead, the code underlying the operations of the GNSS module; although this is a software component whose implementation depends on the type of hardware in use, we believe that the inclusion of this example is of paramount importance, as it demonstrates the approach to be followed when implementing a native C add-on for Node.js. Finally, Listing C.4 is provided, which contains the JavaScript code used by the dongle to estimate the vehicle's emissions based on its specifications and the diagnostic data obtained via the OBD.

LISTING C.1 DONGLE, C FUNCTIONS FOR ISO 15765 DIAGNOSTIC COMMUNICATION

```
//FUNCTION: SendISO15765Message [Basic_Function.c]
//INFO: It requests a specific PID to the ECU via an ISO 15765 message
//INPUT: Socket ID, Address, PID, Supported PIDs, Addressing Mode, CAN Frame
//OUTPUT: Operation Status

int Send_ISO15765Message(int socket, canid_t ADDRESS,   u8 PID, int*
↪  SUPPORTED_PIDs, int ADDRESSING_MODE, struct can_frame* frame){

     //Support Variable
     int nbytes = 0;

     //CAN_ID = ADDRESS + FLAGS
     canid_t CAN_ID;

     //If the PID is supported by the vehicle
     if(SUPPORTED_PIDs[(int)PID]==1){

     //Set CAN_ID according to Addressing Mode
     if(ADDRESSING_MODE==0){
         //Functional Address for CAN 2.0A is 0x7DF (11 bits)
```

```
            if(ADDRESS==DEFAULT) CAN_ID = Functional_11;
            else CAN_ID = ADDRESS;
        }
        else if(ADDRESSING_MODE==1){
            //Functional Address for CAN 2.0B is 0x18DB33F1 (29bits)
            if(ADDRESS==DEFAULT) CAN_ID = Functional_29 | CAN_EFF_FLAG;
            else CAN_ID = ADDRESS | CAN_EFF_FLAG;
        }
        else {
            printf("\nERROR: Addressing Mode not supported\n");
            return(-2);
        }

        //Set CAN message to be sent, PID MODE is set to 0x01
        frame->can_id = CAN_ID;
        frame->can_dlc = 8;
        frame->data[0] = 0x02;
        frame->data[1] = PID_MODE_01;
        frame->data[2] = PID;
        frame->data[3] = PADDING;
        frame->data[4] = PADDING;
        frame->data[5] = PADDING;
        frame->data[6] = PADDING;
        frame->data[7] = PADDING;

        //Send ISO 15765 Message on CAN
        nbytes = send(socket, frame, sizeof(struct can_frame), 0);
        if(nbytes==0) printf("Nothing has been written");
        else if(nbytes==-1){
            perror("Error in writing on socket");
        }
    }

    //If return == 0 -> PID is not supported
    //If return == -1 -> Error in writing on socket
    //If return == -2 -> Addressing Mode not supported
    return nbytes;
}



//FUNCTION: ReceiveISO15765 Message [Basic_Function.c]
//INFO: It retrieves a CAN message from the queue and extracts its content
//INPUT: Socket ID, CAN Frame
//OUTPUT: Operation Status, ISO 15765 Diagnostic Message
//NOTE: The list of decoding formulas can be extended beyond standard PIDs.
// The main source of information on non-standard PIDs of the OEMs,
// accessible only to members and clearly not exhaustive, is
// maintained by the Equipment and Tool Institute (ETI).

```

```
int Receive_ISO15765Message(int socket, struct can_frame* frame, msg* message){

    //Support Variables
    double PID, A,B,C,D,E,F, res = 0;
    int nbytes = 0;

    //Set Buffer Size
    size_t buffer_size = 1000;

    //Extract CAN Frame
    nbytes = recv(socket, frame, buffer_size, 0);

    //Processing of content according to message size
    if(nbytes<0){
       perror("can raw socket read");
    }
    else if (nbytes<sizeof(struct can_frame)){
       fprintf(stderr, "read: incomplete CAN frame\n");
   }
   else {

      //Check: should not be required as a software mask is set on the queue
       if((long)frame->data[1]==MODE01_RESPONSE && frame->can_dlc==8){

          //Extract CAN frame content
          PID = frame->data[2];
          A = (double)frame->data[3];
          B = (double)frame->data[4];
          C = (double)frame->data[5];
          D = (double)frame->data[6];
          E = (double)frame->data[7];
          F = (double)frame->data[8];

          //Decode CAN Frame content based on PID
          switch((int)PID){

             case MONITOR_STATUS:
                res = -1; //Formula to be implemented
                break;

             case FREEZE_DTC:
                res = -1; //Formula to be implemented
                break;

             case FUEL_SYSTEM_STATUS:
                res = -1; //Formula to be implemented
                break;

```

```
            case CALCULATED_ENGINE_LOAD:
               res = A/2.55;
               break;

            case ENGINE_COOLANT_TEMPERATURE:
               res = A-40;
               break;

            case SHORT_TERM_FUEL_TRIM_1:
               res = (A/1.28)-100;
               break;

            case LONG_TERM_FUEL_TRIM_1:
               res = (A/1.28)-100;
               break;

            case SHORT_TERM_FUEL_TRIM_2:
               res = (A/1.28)-100;
               break;

            case LONG_TERM_FUEL_TRIM_2:
               res = (A/1.28)-100;
               break;

            case FUEL_PRESSURE:
               res = 3*A;
               break;

            case INTAKE_MANIFOLD_ABOSLUTE_PRESSURE:
               res = A;
               break;

            case ENGINE_RPM:
               res = ((256*A)+B)/4;
               break;

            case VEHICLE_SPEED:
               res = A;
               break;

            case TIMING_ADVANCE:
               res = (A/2)-64;
               break;

            case INTAKE_AIR_TEMPERATURE:
               res = A-40;
               break;

            case MAF:
               res = ((256*A)+B)/100;
               break;
```

```
            case THROTTLE_POSITION:
               res = (100/255)*A;
               break;

            case COMMANDED_SECONDARY_AIR_STATUS:
               res = -1; //Formula to be implemented
               break;

            case OXYGEN_SENSOR_PRESENT_2:
               res = -1; //Formula to be implemented
               break;
}
            case OXYGEN_SENSOR_1: //OR condition
            case OXYGEN_SENSOR_2:
            case OXYGEN_SENSOR_3:
            case OXYGEN_SENSOR_4:
            case OXYGEN_SENSOR_5:
            case OXYGEN_SENSOR_6:
            case OXYGEN_SENSOR_7:
            case OXYGEN_SENSOR_8:
               res = -1; //Formula to be implemented
               break;

            case OBD_STANDARD:
               res = -1; //Formula to be implemented
               break;

            case OXYGEN_SENSOR_PRESENT_4:
               res = -1; //Formula to be implemented
               break;

            case AUXILIARY_INPUT_STATUS:
               res = -1; //Formula to be implemented
               break;

            case RUN_TIME_SINCE_ENGINE_START:
               res = 256*A + B;
               break;

            case SUPPORTED_PIDS_2:
               res = -1; //Nested Procedure to be implemented
               break;

            //...

            default:
               printf("\nPID not supported...\n");
               res = -1;
               break;
            }
```

```
         //Set Message, if no error has occurred
         if(res!=-1){
            message->PID = PID;
            message->value = res;
         }
      }
   }
   //If nbytes < 0 -> PID is not supported
   //If nbytes < sizeof(struct can_frame) -> Error in writing on socket
   //If nbytes == sizeof(struct can_frame) -> Message available
   return nbytes;
}
```

LISTING C.2 DONGLE, C FUNCTION FOR VIN EXTRACTION WITH ISO 15765

```
//FUNCTION: getVIN [Basic_Function.c]
//INFO: It obtains the VIN using Mode 9 on ISO 15765
//INPUT: Socket ID , Addressing Mode, Address
//OUTPUT: Operation status, VIN

int get_VIN(int* socket, int addressing_mode, canid_t ADDRESS, char* VIN){

     Set CAN ID based on Addressing Mode (CAN_ID = ADDRESS + FLAG)
     //CAN2.0A CAN_ID = 0x7DF;
     //CAN2.0B CAN_ID = 0x18DB33F1 | CAN_EFF_FLAG;


    Set Software Mask on socket via setsockopt() based on Addressing Mode
    //CAN2.0A Filter: ID_RESPONSE_11 0x7E0, ID_MASK_11 0x7F0 (11bits)
    //CAN2.0B Filter: ID_RESPONSE_29 0x18DAF100, ID_MASK_29 0x1FFF0000 (29bits)

     Send Request_Message with CAN_ID
    //dlc:0x08, length:0x02, PID_MODE:0x09, PID:0x02, Others:0x00 [Padding]

     When a Response_Message is received
     //The response must be received within 5 seconds [Timer]

          //If Byte3 is 49 and Byte4 is 2, VIN over Mode 9 is received
          if(Response_Message[2]==0x49 && Response_Message[3]==0x02)

          //If Byte5 is 1, this message is the first of a multi-frame sequence,
          //If Byte1 is 10, at least another message will follow
          if(Response_Message[4]==0x01 && Response_Message[0]==0x10)

               //FLOOR[(Byte 2)/8] is the number of messages that will follow
               sequence_size = floor(Response_Message[1]/8);

```

```
            //Extract Bytes 06,07,08 as the first 3 chars of the VIN
            Set VIN[values];


    //Guard based on sequence_size
    if(the VIN is split into multiple messages)

         Send Flow_Control_Message with CAN_ID
         //dlc: 0x08, Byte0:0x30 [Flow Control], Others: 0x00 [Padding],
         //Byte1,2,3: 0x00 [Options: do not wait/send as soon as possible]

          //The number of cycles is defined by sequence_size
          For(the number of messages that will follow)

              When a FlowControl_Response is received
               //The response must be received within 5 seconds [Timer]

                 //Byte 01 is the sequence number
                 if(FlowControl_Response[0] is the next value of the sequence)

                    //Extract Bytes 02,03,04,05,06,07,08 from Response
                    Set VIN[values];

     //if VIN_status == -1 -> Unable to complete the operation
     //if VIN_status == 1 -> VIN extraction completed successfully
     return VIN_Status
}
```

LISTING C.3 DONGLE, NATIVE C ADD-ON FOR GNSS MODULE CONTROL

```
#include <node_api.h>

// C library headers
#include <stdio.h>
#include <string.h>
#include <stdlib.h>

// Linux headers
#include <fcntl.h> // Contains file controls like O_RDWR
#include <errno.h> // Error integer and strerror() function
#include <termios.h> // Contains POSIX terminal control definitions
#include <unistd.h> // write(), read(), close()

// Functions
#include "Decode_NMEA.h"
```

```
/*********************************************************
//
//INFO:  GPS.c is the entry point that allows to execute AT commands via
//       the Waveshare 4G board. This add-on can be used to turn the GPS
//       on with input == 'on', to turn the GPS off with input == 'off',
//       and to request the GPS position of the vehicle (lat, lon) with
//       input == 'info'
//
//INPUT:     selection {on, off, info} - to select the behaviour of the add-on
//
//OUTPUT:    no output with selection == on
//           no output with selection == off
//           lat, lon with selection == info
//
//FURTHER NOTES:
//        - The add-on is executed synchronously; hence, the Node.js module
//          that invokes it must wait for the add-on to finish its execution;
//        - The add-on is implemented through N-API, which guarantee that
//          the code can be executed on all the supported versions of
//          Node.js, without the need to be recompiled (unlike NAN);
//        - The add-on might be extended in the future to obtain other
//          parameters from the NMEA string like timestamp, date, and altitude.
//
//********************************************************




napi_value gps_info(napi_env env, napi_callback_info info){

    // 0 - Input and Output parameters

    // 0.1 - Get input parameters (1)
    size_t argc = 1; //Number of inputs
    napi_value args[1]; //Structure to save the input in
    napi_status status_input;
    status_input = napi_get_cb_info(env, info, &argc, args, NULL, NULL);
    if (status_input != napi_ok) return NULL;
    char selection [5]; //3 possible inputs so far [max length: 5]
    napi_get_value_string_utf8(env, args[0], selection, 5, NULL);


    // 0.2 - Create Plain object to be returned
    napi_value obj;
    napi_status status_output;
    status_output = napi_create_object(env, &obj);
    if (status_output != napi_ok) return NULL;
```

```
    // 1 - OPEN SERIAL PORT
    // Open the serial port (device path to be set as required)
    // Open for both reading and writing
    int serial_port = open("/dev/ttyUSB2", O_RDWR);

    // Create new termios struct, to control the asynchronous communication
    struct termios tty;

    // Get the parameters of the port and store them in the termios struct
    if(tcgetattr(serial_port, &tty) != 0){
       printf("Error %i from tcgetattr: %s\n", errno, strerror(errno));
        return NULL;
   }


    // 2 - SETUP SERIAL PORT
    tty.c_cflag &= ~PARENB; // Clear parity bit, disabling parity
    tty.c_cflag &= ~CSTOPB; // Clear stop field
    tty.c_cflag &= ~CSIZE; // Clear all bits that set the data size
    tty.c_cflag |= CS8; // 8 bits per byte
    tty.c_cflag &= ~CRTSCTS; // Disable RTS/CTS hardware flow control
    tty.c_cflag |= CREAD | CLOCAL; // Turn on READ & ignore ctrl lines
    tty.c_lflag &= ~ICANON;
    tty.c_lflag &= ~ECHO; //Disable echo
    tty.c_lflag &= ~ECHOE; // Disable erasure
    tty.c_lflag &= ~ECHONL; // Disable new-line echo
    tty.c_lflag &= ~ISIG; // Disable interpretation of INTR,QUIT,SUSP
    tty.c_iflag &= ~(IXON | IXOFF | IXANY); // Turn off s/w flow ctrl
    tty.c_iflag &= ~(IGNBRK|BRKINT|PARMRK|ISTRIP|INLCR|IGNCR|ICRNL);
    tty.c_oflag &= ~OPOST; // Prevent special interpretation of output bytes
    tty.c_oflag &= ~ONLCR; // Prevent conversion of newline to carriage return



    // VMIN > 0, VTIME > 0 [Other combinations determine different behaviours]:
    // Block until either VMIN characters have been received,
    // Or Block until VTIME after first character has elapsed;
    // timeout for VTIME does not start until the first character is received.
    tty.c_cc[VTIME] = 100;
    tty.c_cc[VMIN] = 40;

    // Set in/out baud rate to 9600
    cfsetispeed(&tty, B9600);
    cfsetospeed(&tty, B9600);

    // Save tty settings, also checking for error
    if (tcsetattr(serial_port, TCSANOW, &tty) != 0){
       printf("Error %i from tcsetattr: %s\n", errno, strerror(errno));
    return NULL;
   }
```

```
    // 3 - WRITE TO SERIAL PORT (Request GPS info via AT command)
    unsigned char msg_on[] = "AT+CGPS=1\r\n";
    unsigned char msg_off[] = "AT+CGPS=0\r\n";
    unsigned char msg_info[] = "AT+CGPSINFO\r\n";


    // 3.1 - GPS ON
    if(strcmp(selection,"on")==0){

        // Write on the serial port
        write(serial_port, &msg_on, sizeof(msg_on));

        // Flush serial port to allow proper read later
        sleep(1);
        tcflush(serial_port,TCIOFLUSH);
    }


    // 3.2 - GPS OFF
    else if(strcmp(selection,"off")==0){

        // Write on the serial port
        write(serial_port, &msg_off, sizeof(msg_off));

        // Flush serial port to allow proper read later
        sleep(1);
        tcflush(serial_port,TCIOFLUSH);
    }


    // 3.3 - GPS INFO
    else if(strcmp(selection,"info")==0){

        // Write on the serial port
        write(serial_port, &msg_info, sizeof(msg_info));


        // 3.3.1 - Obtain GPS info: read from serial port

        // Allocate memory for read buffer, set size according to needs
        char read_buf [100];
        memset(&read_buf, '\0', sizeof(read_buf));

        // Read bytes: behaviour depends VMIN and VTIME
        int num_bytes = read(serial_port, &read_buf, sizeof(read_buf));

        // num_bytes is 0 if no bytes were received, -1 if an error occurred
        if (num_bytes <= 0){
```

```
            printf("Error reading: %s", strerror(errno));
            return NULL;
        }


        // 3.3.2 - READ lan,lon from Serial Port via Decode_NMEA.c functions

        //If the GPS is not ready, the functions will return a default value[0]
        //that is required to be managed on a higher level of the software
        int status = Extract_latlon_fromNMEA(read_buf, &lat, &lon);

        //The GPS position can also be retrieved by using:
        //lat = Extract_from_NMEA (read_buf, 1);
        //lon = Extract_from_NMEA (read_buf, 2)


        // 3.3.3 - Return Lat and Lon

        //Convert the obtained values in JavaScript format
        napi_status status;
        napi_value lat_js, lon_js;
        napi_create_double(env, lat, &lat_js);
        napi_create_double(env, lon, &lon_js);

        //Set Properties object (from Input)
        status = napi_set_named_property(env, obj, "lat", lat_js);
        status = napi_set_named_property(env, obj, "lon", lon_js);
        if (status != napi_ok) return NULL;
    }


    // 3.4 - Input non recognised
    else printf("Invalid input value\n");


    // 4 - CLOSE SERIAL PORT
    close(serial_port);


    //Return value
    return obj;
}


//Module registration
//Return the object that allows calling the add-on functions from Javascript
napi_value init(napi_env env, napi_value exports) {

    //Variable used to access the native add-on
    napi_value new_exports;
```

```
    //Set napi_value
    //INPUTS: Environment, Function_Name, Length of Function_Name,
    //        Native_Function, Input_Data, Output_Data
    napi_status status = napi_create_function (env, "", NAPI_AUTO_LENGTH,
    ↪ gps_info, NULL, &new_exports);

    //If the process to create the exe file from the C add-on fails
    if (status != napi_ok) return NULL;

    //Return napi_value
    return new_exports;
}

//INIT the registered module
NAPI_MODULE(NODE_GYP_MODULE_NAME, init)
```

LISTING C.4 DONGLE, JAVASCRIPT FUNCTION FOR ESTIMATING VEHICLE POLLUTION

```
//FUNCTION: estimatePollution [Share_Auxiliary_Functions.js]
//INFO:     It estimates the emissions of the vehicle based on its speed
//INPUT:    Vehicle Speed, Emission Model factors
//OUTPUT:   Estimated Pollutant
//NOTE:     It is based on a formula that estimates one of the four pollutants
//          foreseen, namely CO, HC, NOx, and PM, depending on the factors
//          provided as input, which are in the form of an array of 8 values;
//          each vehicle is in fact associated with a set of 4 distinct arrays,
//          which are determined according to the vehicle specifications during
//          the configuration of the OBD dongle.

function estimatePollution(speed, pollutant_factors) {

    //Value to be returned
    var pollutant

    //Adjusted speed
    var s

    //Adjust Vehicle Speed, if required:
    //Set value to min (5) and max (140) if speed is out of range
    if(speed<5)s = 5
    else if(speed>140)s = 140
    else s = speed

```

```
    //Calculate powers of speed
    //The multiplication operator appears to be faster than Math.pow
    var s2 = s * s
    var s3 = s2 * s
    var s4 = s3 * s
    var s5 = s4 * s
    var s6 = s5 * s


    //Estimate Emission
    //pollutant = (k * (a + b*s+ c*s2 + d*s3 + e*s4 + f*s5 + g*s6))/s
    pollutant = (pollution_factors[7] * ((pollution_factors[0]) +
    ↪  (pollution_factors[1]*x) + (pollution_factors[2]*x2) +
    ↪  (pollution_factors[3]*x3) + (pollution_factors[4]*x4) +
    ↪  (pollution_factors[5]*x5) + (pollution_
       factors[6]*x6)))/x

    //Return Estimated Emission
    return pollutant
}
```

Appendix D
User Interfaces Code

The purpose of Appendix D is to provide an in-depth examination of the software supporting the operations of the two user interfaces of the CVP, already discussed in Chapter 7. Since both the mobile and the web applications are relatively standard in terms of technology stack and software architecture, we refrain from listing their entire source code, preferring to focus on the software elements that set these components apart from the user interfaces used by conventional FMS. Provided below is Listing D.1, showing the code of the *MQTTHelper* class that the mobile application uses to interact with the broker; it is followed by Listing D.2, which illustrates the structure of the *CallMonitoring* module, responsible for the foreground service through which the smartphone can receive MQTT messages even when the mobile application is not active on screen.

The remaining listings illustrate the queries that the web app uses to retrieve data from the database. Listings D.3 presents the queries behind the monitoring of fleet diagnostics, demonstrating how PostGIS is used to filter database records within a specified geographic area; Listings D.4 and D.5 demonstrate the platform's ability to jointly query vehicle and context data, which enables functionalities that are not achievable with the current fleet management system employed by the NHS. Each listing is accompanied by a series of screenshots of the corresponding section of the UI, allowing readers to visually associate the operations performed in the CVP's back-end with the information that is displayed on the web page.

LISTING D.1 MOBILE APP, JAVA CLASS FOR MQTT COMMUNICATION

```
//It manages the interactions of the mobile app with the CVP over MQTT
//It extends Observable to notify the receipt of a message to its Observers

public class MQTTHelper extends Observable {

    //Support Variables ***************************

    //Broker
    private String broker;

    //Topic
    private String topic;

    //QoS
    private int qos;

```

```
    //MQTT Message
    private String message;

    //Client Connection Option
    private MqttConnectOptions options = new MqttConnectOptions();


    //Constructors *******************************

    //Empty Constructor
    public MQTTHelper(){}


    //Getter and Setter ************************

    //Get Broker
    public String getBroker(){ return broker; }

    //Set Broker
    public void setBroker(String broker){ this.broker = broker; }

    //Get Topic
    public String getTopic(){ return topic; }

    //Set Topic
    public void setTopic(String topic){ this.topic = topic; }

    //Get QoS
    public int getQos() { return qos; }

    //Set QoS
    public void setQos(int qos){ this.qos = qos; }

    //Get Message
    public String getMessage(){ return message; }

    //Set Topic
    public void setMessage(String message){ this.message = message; }


    //Functions********************************

    //FUNCTION 1 - Connect to MQTT Broker
    public MqttAndroidClient Connect(Context context, String id){

        //Client ID
        String clientId;

```

```
        //Autogenerate Client ID, if ID is not provided as input
        if(id.equals(null)) clientId = MqttClient.generateClientId();

        //Otherwise, obtain Client ID from input
        else clientId = id;

        //Create MQTT Client
        MqttAndroidClient client = new MqttAndroidClient(context, broker, clientId);

        //Set Persistent Connection
        options.setCleanSession(false);

        //Set Connection Timeout, default value is 1
        options.setKeepAliveInterval(10);

        //Set Automatic Reconnection, to be used when connection fails
        options.setAutomaticReconnect(true);


    try {
        //Connect
        IMqttToken token = client.connect(options);

        //Set Callbacks
        token.setActionCallback(new IMqttActionListener() {

            @Override
            public void onSuccess(IMqttToken asyncActionToken) {
                Log.d("MQTT", "onSuccess");
            }

            @Override
            public void onFailure(IMqttToken asyncActionToken, Throwable exception) {
                //It should reconnect with exponential backoff
                Log.d("MQTT", "onFailure");
            }
        });


        //Set Extended Callbacks
        client.setCallback(new MqttCallbackExtended() {

            @Override
            public void connectComplete(boolean b, String s) {

            try {
                IMqttToken subToken = client.subscribe(topic, qos);
                subToken.setActionCallback(new IMqttActionListener() {

```

```
                    @Override
                    public void onSuccess(IMqttToken asyncActionToken){
                        Log.d("MQTT", "Subscription to " + topic);
                    }

                    @Override
                    public void onFailure(IMqttToken asyncActionToken,
                    ↪  Throwable exception) {
                       Log.d("MQTT", "Subscription failed");
                    }
                });

            } catch (MqttException e) {
                e.printStackTrace();
            }
        }

        @Override
        public void connectionLost(Throwable throwable) {
            Log.d("MQTT", "MQTT client connection lost");
        }

         @Override
         public void deliveryComplete(IMqttDeliveryToken
         ↪  iMqttDeliveryToken) {
             Log.d("MQTT", "Delivery completed - callback");
         }


         @Override
         public void messageArrived(String topic, MqttMessage
         ↪  mqttMessage) throws Exception {

            //MQTT Message to be shared with CallMonitoring (as JSON)
            MQTTHelper mqtthelper = new MQTTHelper();
            mqtthelper.setMessage(mqttMessage.toString());

            //Notify Changes to CallMonitoring
            setChanged();
            notifyObservers(mqtthelper);
            }
        });

    } catch (MqttException e) {
        e.printStackTrace();
    }

    //Return client to allow subsequent operations on it
    return client;
```

```
    //FUNCTION 2 - Disconnect from MQTT Broker
    public void disconnect(Context context, MqttAndroidClient client){

        try {
            //Disconnect
            IMqttToken disconToken = client.disconnect();v

            //Set token callbacks (onSuccess, onFailure)
            disconToken.setActionCallback(new IMqttActionListener() {

                @Override
                public void onSuccess(IMqttToken asyncActionToken) {
                    Log.d("MQTT", "Client disconnected - callback");

                    //Disable automatic reconnect, to avoid immediate
                    //reconnection after explicit user request to disconnect
                    options.setAutomaticReconnect(false);
                }

                @Override
                public void onFailure(IMqttToken asyncActionToken, Throwable exception) {
                    Log.d("MQTT", "Impossible to disconnect client");
                }
            });
        }
        catch(MqttException e){
            e.printStackTrace();
        }
    }

    //FUNCTION 3 - Publish Message on MQTT Topic
    public void publish(Context context, MqttAndroidClient client, String payload){

        //Set Message from input
        MqttMessage message = new MqttMessage(payload.getBytes());

        //Set MQTT Communication settings
        message.setQos(qos);
        message.setRetained(false);

        //Publish
        try {
            client.publish(topic, message);
        } catch (MqttException e) {
            e.printStackTrace();
        }
    }
}
```

LISTING D.2 MOBILE APP, JAVA CLASS FOR FOREGROUND CALL MONITORING

```
//Foreground service used to receive notifications regarding a change in the
//operational status of the vehicle; it implements Observer to receive
//automatic notifications about the opening and closing of a call following
//the receipt of an MQTT message by the observed MQTTHelper instance.

public class CallMonitoring extends Service implements Observer {


    //Support Variables**************************

    //Application Context
    private Context context;

    //Notification Channel
    public static final String CHANNEL_ID = "CallMonitoringChannel";

    //Notification Manager
    private NotificationManagerCompat notification Manager;

    //Notification
    private Notification notification;

    //Notification ID, to enable the notification to be updated
    private int notifyID = 1;

    //Variable to manage Notification title and Notification text
    private String vehicle_info, call_info;

    //Intents
    private Intent notifyIntent;
    private PendingIntent pendingIntent;


    //Communication variables ******************

    //Binder to allow a Call instance to use this service via public methods
    private final IBinder binder = new LocalBinder();

    //MQTT Helper instance
    private MQTTHelper mqttHelper = new MQTTHelper();

    //MQTT client that is returned by MQTTHelper connect()
    private MqttAndroidClient client = null;

    //ID Root, to which staff_id is added to obtain client_ID.
    private String client_id_root = "Staff_";
    private String client_id = "";
```

```
    //The address of the MQTT broker must be defined here
    private String broker;

    //Topic root, to which vehicle_id is added to obtain the MQTT_topic
    private String topic_root = "VEICOLNET/vehicle/";
    private String topic = "";

    //MQTT QoS: 2, the MQTT message should be received exactly once
    private int qos = 2;

    //Responsible for sending data to the UI thread
    private final Handler mHandler = new Handler(Looper.getMainLooper());

    //String to Identify the SendBroadcast intent
    public final static String CALL_ACTION = "CALL_MONITORING";


    //Class Functions ****************************

    @Override
    public void onCreate() {

        //Set Context
        context = this.getApplicationContext();

        //Create Observer
        createObserver(mqttHelper);
        super.onCreate();
    }

    @Override
    public IBinder onBind(Intent intent) {
        return binder;
    }

    @Override
    public void onDestroy() {

        //Unsubscribe from "this_vehicle" topic and Disconnect from MQTT Broker
        mqttHelper.disconnect(context, client);

        //Delete Observer
        deleteObserver(mqttHelper);
        super.onDestroy();
    }

    @Override
    public int onStartCommand(Intent intent, int flags, int startId) {

```

```
        //Set notification title and content based on input (vehicle_id)
        int vehicle_id = intent.getIntExtra("vehicle_id", 0);
        vehicle_info = "Call monitoring on vehicle " + vehicle_id;
        call_info = "No open call";

        //Get Staff_id, to be used to setup Cliend ID
        int staff_id = intent.getIntExtra("staff_id", 0);

        //Create Notification Channel
        createNotificationChannel();
        notifyIntent = new Intent(this, MainActivity.class);

        //Set Notification Manager
        notificationManager = NotificationManagerCompat.from(this);

        //Set Notification Parameters
        notification = new NotificationCompat.Builder (this, CHANNEL_ID)
            .setContentTitle(vehicle_info)
            .setContentText(call_info)
            .setSmallIcon(R.drawable.info)
            .setContentIntent(pendingIntent)
            .build();

        //Start Foreground Service
        startForeground(1, notification);

        //Connect to Broker and Subscribe to "this_vehicle", if not yet done
        if(client==null){

            //Set MQTT topic based on vehicle_id
            topic = topic_root + vehicle_id;

            //Create MQTTHelper
            mqttHelper.setBroker(broker);
            mqttHelper.setTopic(topic);
            mqttHelper.setQos(qos);

            //Create ClientID
            client_id = client_id_root + staff_id;

            //If no cliend_id is provided, it will be auto-generated
            client = mqttHelper.Connect(context, client_id);
        }

        //START_STICKY: if the OS kills this service without
        //explicit user request, it will be restarted
        return START_STICKY;
    }
```

```java
    //Observer Functions *************************

    //Create Observer
    public void createObserver(Observable o) { o.addObserver(this); }

    //Delete Observer
    public void deleteObserver(Observable o) { o.deleteObserver(this); }


    //Public Functions **************************

    //FUNCTION 1 - Get Foreground Service
    //It returns an instance of this service, so that the call class (client)
    //can invoke public methods on it [enabling call->service communication]
    public class LocalBinder extends Binder {
        public CallMonitoring getService() {
            return CallMonitoring.this;
    }   }

    //FUNCTION 2 - Update Service Notification
    //It sets the text and the content of the notification based on input
    public void updateServiceNotification(int id, int emergency_level){

        //Initialize Notification Content based on input parameters
        if(id==0) call_info = "No open call";
        else call_info = "Current call: " +id+ "(Level " +emergency_level+")";

        //Update Notification
        NotificationCompat.Builder builder = new
        ↪ NotificationCompat.Builder(this, CHANNEL_ID);
        builder.setContentText(call_info)
               .setContentTitle(vehicle_info)
               .setSmallIcon(R.drawable.info)
               .setContentIntent(pendingIntent);
        notificationManager.notify(notifyID, builder.build());
    }

    //FUNCTION 3 - Notify Call Closure
    //It delegates to MQTTHelper the task of notifying the closure of the
    //current call by publishing an MQTT message on the topic of the vehicle
    public void notifyCloseCall() {

        //Create Object to be sent over MQTT
        JSONObject closed_call_info = new JSONObject();

        //Add call_id
        try {
            closed_call_info.put("call_id", 0);
        } catch (JSONException e) {
            e.printStackTrace();
        }
```

```
        //Convert JSON object to String
        String message = closed_call_info.toString();

        //Publish call_id = 0 on "this_vehicle" topic
        mqttHelper.publish(context, client, message);
    }


    //Private Functions **************************

    //FUNCTION 5 - Create Notification Channel
    private void createNotificationChannel() {
        if (Build.VERSION.SDK_INT >= Build.VERSION_CODES.O){
           NotificationChannel serviceChannel = new
           ↪ NotificationChannel(CHANNEL_ID, "Call Monitoring Service
           ↪ Channel", NotificationManager.IMPORTANCE_LOW);
           NotificationManager manager =
           ↪ getSystemService(NotificationManager.class);
           manager.createNotificationChannel(serviceChannel);
        }
    }

    //Update Observers ****************************
    //It notifies a change in the operational status of the vehicle, either
    //the opening of a call or the closure of the current one. This information
    //is shared to main activity of the mobile application, via the call class,
    //through a broadcast receiver.

    @Override
    public void update(Observable observable, Object o) {

        //Use Main looper to send data back to UI Thread
        mHandler.post(new Runnable() {

        @SuppressLint("WrongConstant")
        @Override
        public void run() {

            //When the state of the observed instance changes
            if (o != null){

                //If the state of the MQTTHelper instance has changed
                if (o instanceof MQTTHelper) {

                    //Get response in a local MQTTHelper variable
                    MQTTHelper local_mqttHelper = new MQTTHelper();
                    local_mqttHelper = (MQTTHelper) o;

```

```
                        //Extract message
                        String string_message = local_mqttHelper.getMessage();
                        int call_id = 0;
                        int emergency_level = 0;

                        //Extract relevant fields from message
                        JSONObject obj_message = null;

                        //EmergencyLevel is available when a call is opened;
                        //When a call is closed only call_id = 0 is published
                        try {
                            obj_message = new JSONObject(string_message);
                            call_id = obj_message.getInt("call_id");
                            if(call_id!=0) emergency_level =
                            ↪  obj_message.getInt("priority");
                        } catch (JSONException e) {
                            e.printStackTrace();
                        }

                        //Update Notification
                        updateServiceNotification(call_id, emergency_level);

                        //Wrap message and send it via Broadcast
                        Intent intent = new Intent();
                        intent.setAction(CALL_ACTION);
                        intent.putExtra("msg", local_mqttHelper.getMessage());
                        sendBroadcast(intent);


                        //Play sound according to new call status (closed/open)
                        MediaPlayer mediaPlayer;
                        if(call_id==0) mediaPlayer =
                        ↪  MediaPlayer.create(context,
                        ↪  R.raw.closed_call_sound);
                        else mediaPlayer = MediaPlayer.create(context,
                        ↪  R.raw.open_call_sound);
                        mediaPlayer.start();
                    }
                }
            } //run
        }); //mHandler
    } //update

}
```

LISTING D.3 WEB APP, SQL QUERIES FOR DIAGNOSTICS MONITORING

```
-- FLEET DIAGNOSTICS MONITORING


-- FUNCTION 1: getLatestMeasurementsByArea [Multi-Vehicle Monitoring]
-- INPUT:  latSW, lonSW, latNE, lonNE, endTime [set to currentTime - 10s]
-- OUTPUT: Latest measurement for each vehicle in the area [List]

-- INFO:
-- It obtains a list containing 1 measurement for each of the vehicles located
-- in a certain geographical area, aligned to the most recent valid timestamp
-- within the last 10 seconds. It employs two queries; the first query obtains
-- a sorted list of the different timestamps that characterise the measurements
-- that were produced in the area in the last 10 seconds; the second query uses
-- the second of the most recent timestamps to retrieve a measurement for each
-- of the vehicles in that area; this ensures that all measurements for the
-- chosen timestamp have been received. Both queries use ST_Intersects provided
-- by PostGIS to filter the database data; since the area is a a GIS polygon,
-- namely a closed linear ring, and more precisely a square, five points are
-- needed to represent it, with the last point coinciding with the first one.

-- Query 1:
   SELECT DISTINCT measurement_timestamp
   FROM measurement
   WHERE (measurement_timestamp >= 'endTime')
   AND ST_Intersects(ST_Polygon('LINESTRING(lonNE latNE, lonNE latSW, lonSW
   ↪  latSW, lonSW latNE, lonNE latNE)'::geometry, 4326), gps_position)
   ORDER BY measurement_timestamp DESC;

-- Query 2:
   SELECT measurement_timestamp, dongle_id, vehicle_id, lat,lon, co, engine_rpm
   FROM measurement
   WHERE (measurement_timestamp = 'c_timestamp')
   AND ST_Intersects(ST_Polygon('LINESTRING(lonNE latNE, lonNE latSW, lonSW
   ↪  latSW, lonSW latNE, lonNE latNE)'::geometry, 4326), gps_position)
   ORDER BY vehicle_id;


-- FUNCTION 2: getLatestMeasurementsByVehicleID [Single-Vehicle Monitoring]
-- INPUT:  chosen_vehicleID, endTime [set to currentTime - 10s]
-- OUTPUT: Latest vehicle measurement

-- INFO:
-- It obtains the latest measurement of the vehicle with the given ID. It only
-- retrieves records with a valid GPS position produced in the last 10 seconds.

--Query:
   Select * from measurement
   WHERE (vehicle_id = 'chosen_vehicleID')
   AND (measurement_timestamp >= 'endTime')
   AND (lat != 0) AND (lon != 0)
   ORDER BY measurement_timestamp DESC;
```

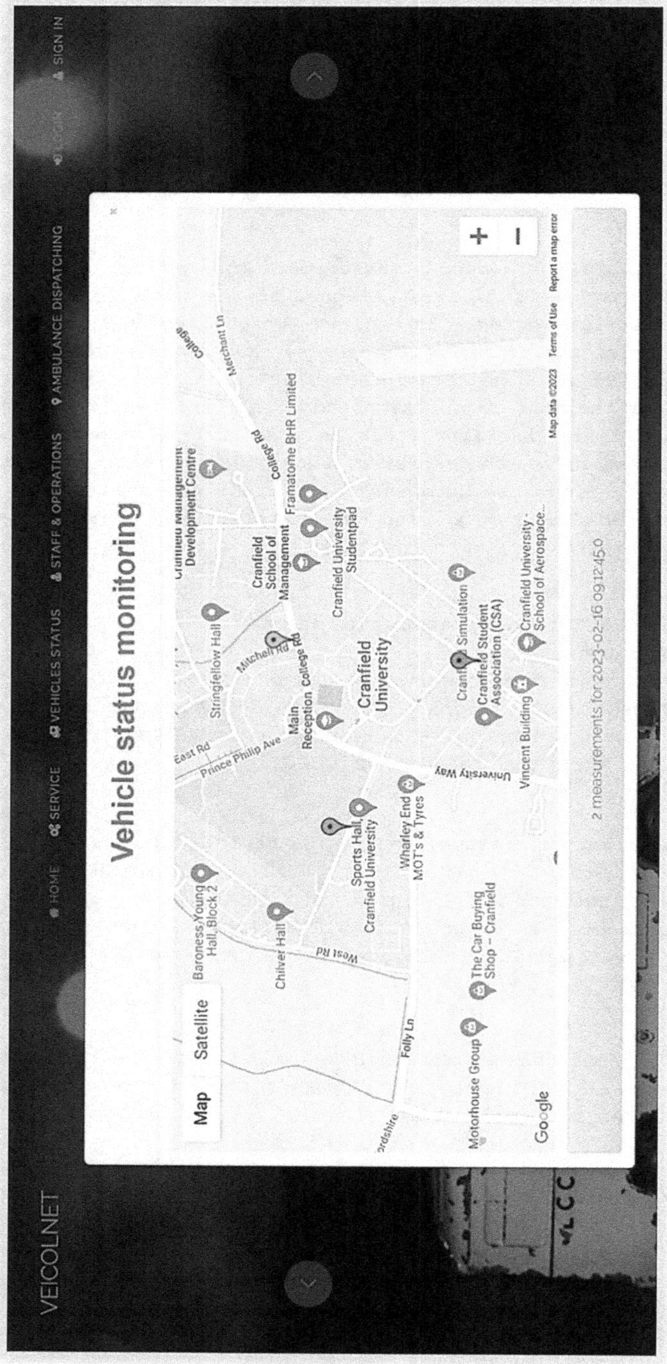

FIGURE D.1 Web app, UI details: Multi-vehicle diagnostics monitoring.

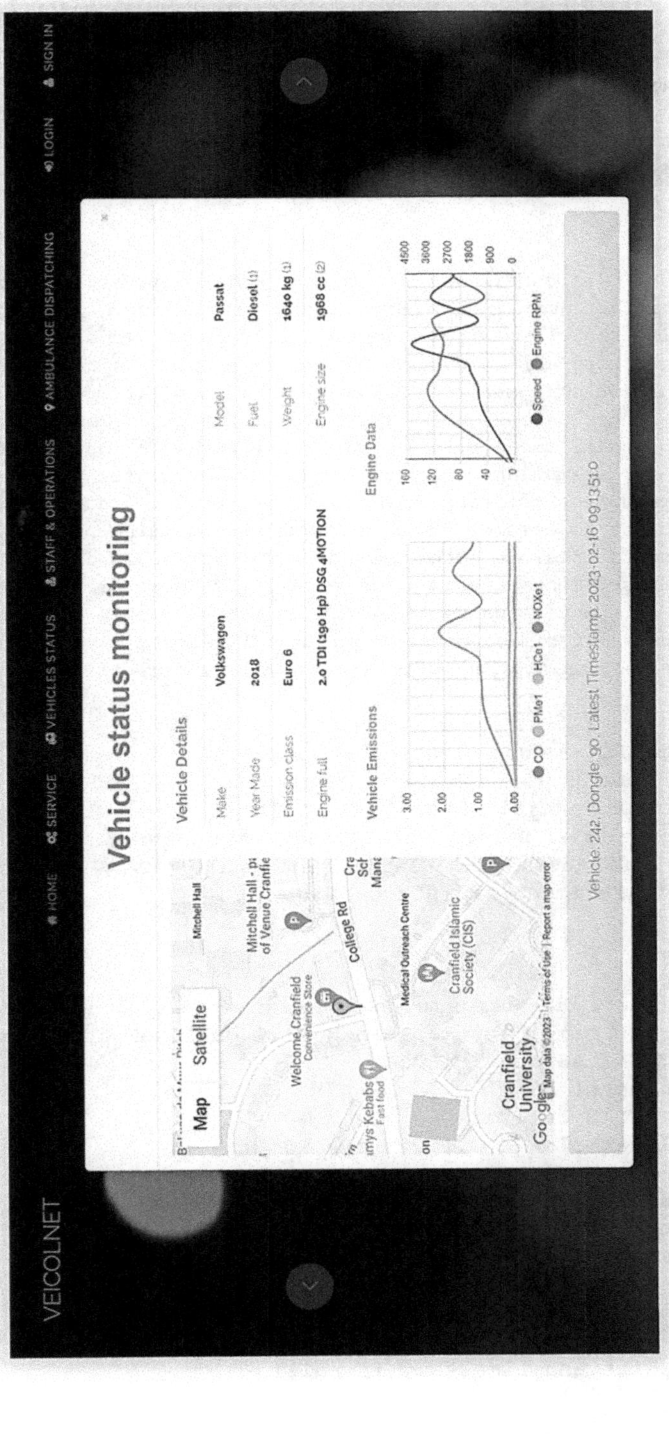

FIGURE D.2 Web app, UI details: Single-vehicle diagnostics monitoring.

LISTING D.4 WEB APP, SQL QUERIES FOR OPERATIONS MONITORING

```
-- FLEET OPERATIONS MONITORING

-- FUNCTION 1: getLatestOperationsByArea [Multi-Vehicle Monitoring]
-- INPUT:  latSW, lonSW, latNE, lonNE, endTime [set to currentTime - 10s]
-- OUTPUT: Operational status for each vehicle in the area [Operation List]


-- INFO:
-- It obtains a list containing information on the operational status of each
-- of the vehicles located in a certain geographical area, aligned to the most
-- recent valid timestamp within the last 10 seconds. This function relies on
-- two database queries, which serve the same purpose as the queries used for
-- monitoring fleet diagnostics. The element that differentiates the two
-- functions is the dataset that is obtained with the second query, which, in
-- the case of this function, does not simply retrieve the latest measurement
-- for each of the ambulances in the area, but integrates these records with
-- the context data associated with it. This query, in fact, relies on the
-- JOIN construct to combine the records of the measurement table with those
-- of calls and shifts; in both cases an OUTER LEFT JOIN is used, to ensure
-- that the measurement data is returned even if there are no active calls
-- or shifts on the various vehicles. In terms of context data, this function
-- returns the start timestamp of the last call served by the vehicle and its
-- status, as well as the number of ambulance operators that are active on it.

--Query 1:
  SELECT DISTINCT measurement_timestamp
  FROM measurement
  WHERE (measurement_timestamp >= 'endTime')
  AND ST_Intersects(ST_Polygon('LINESTRING(lonNE latNE, lonNE latSW, lonSW
  ↪  latSW, lonSW latNE, lonNE latNE)'::geometry, 4326), gps_position)
  ORDER BY measurement_timestamp DESC;

--Query 2:
  SELECT
     m.measurement_timestamp, m.dongle_id, m.vehicle_id, m.lat, m.lon,
     call.start_timestamp as start_call, call.service AS status,
     shift.COUNT AS staff_number
  FROM measurement AS m
  LEFT OUTER JOIN call
     ON call.vehicle_id = m.vehicle_id
     AND m.measurement_timestamp >= call.start_timestamp
     AND (call.end_timestamp >= m.measurement_timestamp
        OR call.end_timestamp is null)
  LEFT OUTER JOIN shift
     ON shift.vehicle_id = m.vehicle_id
     AND m.measurement_timestamp >= shift.start_timestamp
     AND (shift.end_timestamp >= m.measurement_timestamp
        OR shift.end_timestamp is null)
```

```
    WHERE (m.measurement_timestamp = 'c_timpestamp')
    AND ST_Intersects(ST_Polygon('LINESTRING(lonNE latNE, lonNE latSW, lonSW
    ↪  latSW, lonSW latNE, lonNE latNE)'::geometry, 4326), m.gps_position)
    GROUP BY m.measurement_timestamp, m.vehicle_id, m.dongle_id, m.lat, m.lon,
       call.start_timestamp, call.service
    ORDER BY m.vehicle_id;


-- FUNCTION 2: getLatestOperationByVehicleID [Single-Vehicle Monitoring]
-- INPUT:  chosen_vehicleID, endTime [set to currentTime - 10s]
-- OUTPUT: Information on the operational status of the vehicle [Operation]

-- INFO:
-- It obtains the operational status of the vehicle with the given ID. It only
-- retrieves records with a valid GPS position produced in the last 10 seconds.

-- Query:
    SELECT
       m.measurement_timestamp, m.dongle_id, m.vehicle_id, m.lat, m.lon,
       call.start_timestamp AS start_call, call.service AS status,
       shift.COUNT AS staff_number
    FROM measurement AS m
    LEFT OUTER JOIN call
       ON call.vehicle_id = m.vehicle_id
       AND m.measurement_timestamp >= call.start_timestamp
       AND (call.end_timestamp >= m.measurement_timestamp
          OR call.end_timestamp is null)
    LEFT OUTER JOIN shift
       ON shift.vehicle_id = m.vehicle_id
       AND m.measurement_timestamp >= shift.start_timestamp
       AND (shift.end_timestamp >= m.measurement_timestamp
          OR shift.end_timestamp is null)
    WHERE (m.vehicle_id = 'chosen_vehicleID')
       AND (m.measurement_timestamp >= 'endTime')
       AND (m.lat != 0) AND (m.lon != 0)
    GROUP by m.measurement_timestamp, m.vehicle_id, m.dongle_id, m.lat, m.lon
       call.start_timestamp, call.service
    ORDER BY m.measurement_timestamp DESC;
```

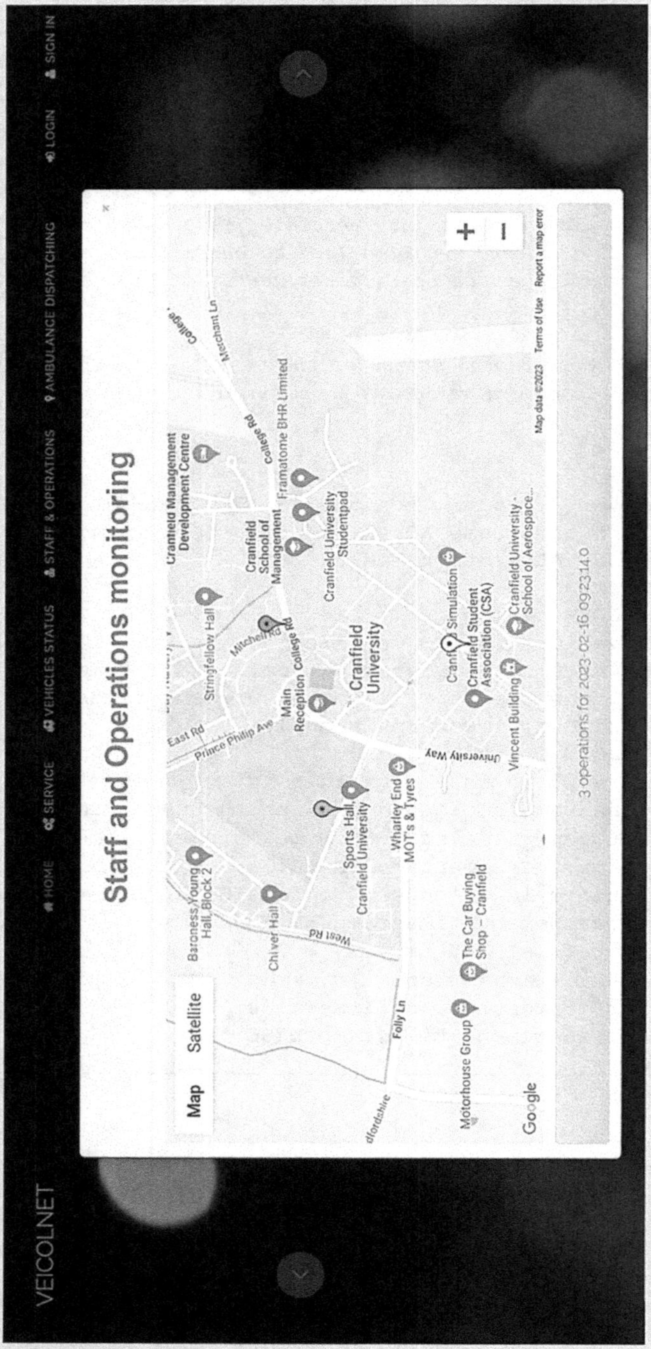

FIGURE D.3 Web app, UI details: Multi-vehicle operations monitoring.

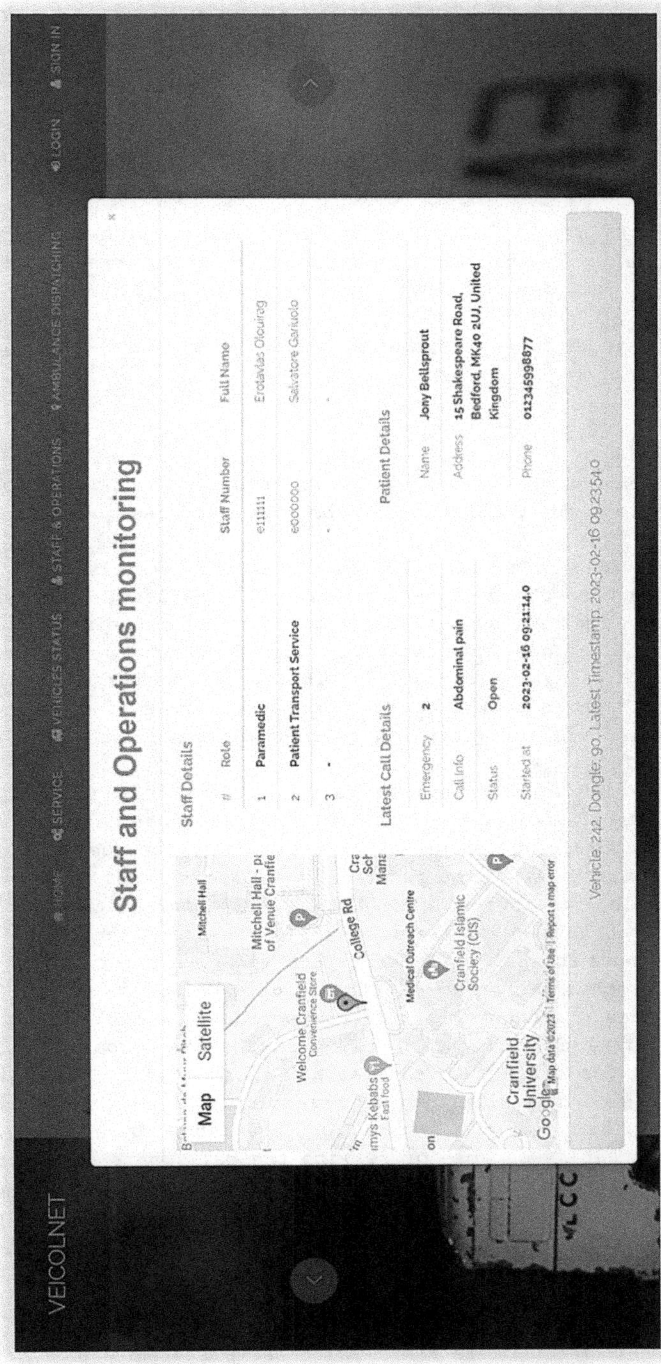

FIGURE D.4 Web app, UI details: Single-vehicle operations monitoring.

LISTING D.5 WEB APP, JAVA FUNCTIONS WITH BUILT-IN SQL FOR VEHICLE DISPATCH

```
-- AMBULANCE DISPATCH


-- FUNCTION 1: getMostSuitableVehicle [Vehicle Selection]
--
-- INPUT: callID [call information is already in the database]
-- disptatchRange [in meters], endTime [set to currentTime - 10s]
-- numVehicles [number of items to be returned, currenlty set to 5]
--
-- OUTPUT: ID of the most suitable vehicle to serve the given call
--
--
-- INFO:
-- It obtains the ID of the most suitable vehicle to handle a certain call.
-- It is based on the combined use of a database query and the Google Maps'
-- DistanceMatrix service, which uses the result of the query as input.
--
-- The query provides information on the 5 available vehicles with the shortest
-- linear distance to the destination that are capable of clinically handling
-- the emergency. It returns the IDs of the vehicles, their GPS position, and
-- the address of the destination to be reached, which is retrieved from the
-- call table via callID. Although the number of vehicles is limited to 5, no
-- LIMIT clause appears in the query, since the number of elements to be
-- returned is set via the appropriate Hibernate function in the Java code.
--
-- This query performs the selection of the five vehicles to be provided to
-- the DistanceMatrix service by interrogating a dataset that is obtained by
-- appropriately combining vehicle and context data. The query relies, in fact,
-- on a LEFT JOIN to combine the records of the measurement table with those of
-- the calls table, in order to obtain the operational status of the vehicles,
-- even those that have never served a call; the use of an EXISTS predicate
-- allows the query to assess whether these vehicles are currently free or
-- engaged in another emergency. The query makes also use of an INNER JOIN
-- between the measurement table and the records of a table obtained by
-- combining shift and staff data, which contains the number of operators that
-- are active on the different vehicles and their overall skill level. This
-- information is used by the query to select only those vehicles that, within
-- the maximum distance to the destination, are manned by a crew with at least
-- one operator that is capable of serving the call; this evaluation is based
-- on a simple comparison between the skill level of the ambulance crews and
-- the level of urgency of the emergency to be handled. To calculate the linear
-- distance between the vehicles and the destination of the call, the query
-- resorts to the ST_DistanceSpheroid function of PostGIS, which uses the
-- World Geodetic System (WGS84) as the reference coordinate system.
--
-- Once the five most suitable vehicles to serve the emergency have been
-- selected based on the information that is available in the database, the
-- dispatching function submits their GPS position and the address of the
```

```
-- destination to be reached to the DistanceMatrix API. The latter returns
-- the shortest driving distance between each vehicle and the destination,
-- as well as the corresponding driving time, which is calculated based on
-- both historical and real-time traffic conditions ('best guess' approach).
-- The function then selects the vehicle with the shortest driving time to
-- reach the destination, whose ID is returned as output.


-- Query [HQL]:
-- Stored in a Java string and executed by the relevant Hibernate function,
-- after setting the maximum number of items to be returned to 5 (numVehicles).

    SELECT
        DISTINCT(m.vehicle_id), timestamp_info.max_timestamp, staff_info.level,
        ST_DistanceSpheroid(m.gps_position, destination.gps_position,
        ↪  'SPHEROID[\"WGS 84\",6378137,298.257223563]') as distance,
        m.lat, m.lon, destination.address
    FROM measurement m
    INNER JOIN call destination
        ON destination.call_id = 'callID'
    INNER JOIN calltype emergency
        ON emergency.calltype_id = destination.calltype_id
    INNER JOIN
        (SELECT
            sh.vehicle_id, COUNT(st.vehicle_role_id) AS staff_number,
            MAX(st.vehicle_role_id) AS level
        FROM shift sh
            INNER JOIN staff st
            ON sh.staff_id = st.staff_id
        WHERE sh.end_timestamp is null
        GROUP BY sh.vehicle_id) staff_info
        ON staff_info.vehicle_id = m.vehicle_id
    LEFT JOIN call cc
        ON m.vehicle_id = cc.vehicle_id
    INNER JOIN
        (SELECT MAX(measurement_timestamp) AS max_t, vehicle_id
        from measurement
        WHERE measurement_timestamp>= 'endTime'
        GROUP BY vehicle_id) timestamp_info
        ON timestamp_info.max_t = m.measurement_timestamp
    WHERE staff_info.level >= emergency.priority
        AND NOT EXISTS (SELECT 1 FROM call WHERE vehicle_id = m.vehicle_id
            AND end_timestamp is null)
        AND ST_DistanceSpheroid(m.gps_position, destination.gps_position,
        ↪  'SPHEROID[\"WGS 84\",6378137,298.257223563]') <= 'dispatchRange'
        AND m.lat != 0 AND m.lon != 0
    ORDER BY distance ASC, timestamp_info.max_t DESC, staff_info.level ASC;
```

```
-- DistanceMatrix API call [Java Pseudocode]:
-- It describes the operations that the Java function performs in terms of
-- interaction with the DistanceMatrix service and the processing that follows.

   Set Origin using the GPS position of the 5 vehicles obtained with the query;
   Set Destination using the address of the call obtained with the query;
   Set the URL of the API Request, using Origin, Destination, and the API key;
   Send the HTTP Request;
   When the HTTP Response is received:
      Select the vehicle with the shortest driving time to the destination;
      Return the ID of the selected vehicle;


-- FUNCTION 2: getVehicleInfo [Selected Vehicle Monitoring]
--
-- INPUT:  chosen_vehicleID, callID, endTime [Set to CurrentTime - 10s]
--
-- OUTPUT: VehicletoDispatch [status and driving info to destination]
--
-- INFO:
-- It is used to obtain information on the status of the ambulance chosen to
-- serve a certain call, both before the vehicle is activated and after it
-- has been dispatched. It is based on the combined use of a database query,
-- employed to retrieve a particular combination of vehicle and context data,
-- and the DistanceMatrix service of Google Maps, which provides information
-- on the driving time required for the vehicle to reach its destination.
--
-- The query returns the GPS position of the vehicle, its operational status,
-- the number of ambulance operators active on it, and their overall skill
-- level; it also returns some vehicle specifications, namely its make and
-- model, as well as details of any call that is being served, including an
-- estimate of the linear distance between the vehicle and its destination.
-- This information is retrieved from the database based on the latest
-- timestamp produced within the last 10 seconds with a valid GPS position.
--
-- The function submits the GPS position of the vehicle and the address of
-- its destination, both obtained by querying the database, to DistanceMatrix.
-- The latter returns the address corresponding to the location of the vehicle
-- (reverse geocoding) and the driving time required for the ambulance to
-- reach the destination. The function adds this information to that obtained
-- from the query, which is returned as output.



-- Query [HQL]:
-- Stored in a Java string and executed by the relevant Hibernate function.

   SELECT
      DISTINCT(m.vehicle_id), v.make, v.model,
      timestamp_info.max_timestamp,
```

```
        m.dongle_id, m.lat, m.lon,
        cc.call_id AS current_call_id,
        vehicle_info.staff_number, vehicle_info.level,
        ST_DistanceSpheroid(m.gps_position, destination.gps_position,
        ↪  'SPHEROID[\"WGS 84\",6378137,298.257223563]') AS distance,
        destination.address AS destination_address, destination.service AS status
    FROM measurement m
    INNER JOIN call destination
        ON destination.call_id = 'callID'
    INNER JOIN vehicle v
        ON m.vehicle_id = v.vehicle_id
    INNER JOIN
        (SELECT
            sh.vehicle_id, COUNT(st.vehicle_role_id) as staff_number,
            MAX(st.vehicle_role_id) AS level
        FROM shift sh
            INNER JOIN staff st
            ON sh.staff_id = st.staff_id
        WHERE sh.end_timestamp is null
        GROUP BY sh.vehicle_id) vehicle_info
        ON vehicle_info.vehicle_id = m.vehicle_id
    INNER JOIN
        (SELECT MAX(measurement_timestamp) AS max_timestamp, vehicle_id
        FROM measurement
        WHERE measurement_timestamp >= 'endTime'
        GROUP BY vehicle_id) timestamp_info
        ON timestamp_info.max_timestamp = m.measurement_timestamp
    LEFT JOIN call cc
        ON m.vehicle_id = cc.vehicle_id AND cc.end_timestamp is null
    WHERE m.vehicle_id = 'chosen_vehicleID'
    AND m.lat != 0 AND m.lon != 0;



-- DistanceMatrix API call [Java Pseudocode]:
-- It describes the operations that the Java function performs in terms of
-- interaction with the DistanceMatrix service and the processing that follows.

    Set Origin using the GPS position of the vehicle obtained with the query;
    Set Destination using the address of the call obtained with the query;
    Set the URL of the API Request, using Origin, Destination, and the API key;
    Send the HTTP Request;
    When the HTTP Response is received:
        Add driving information to the query result;
        Return the overall dataset [VehicleToDispatch];
```

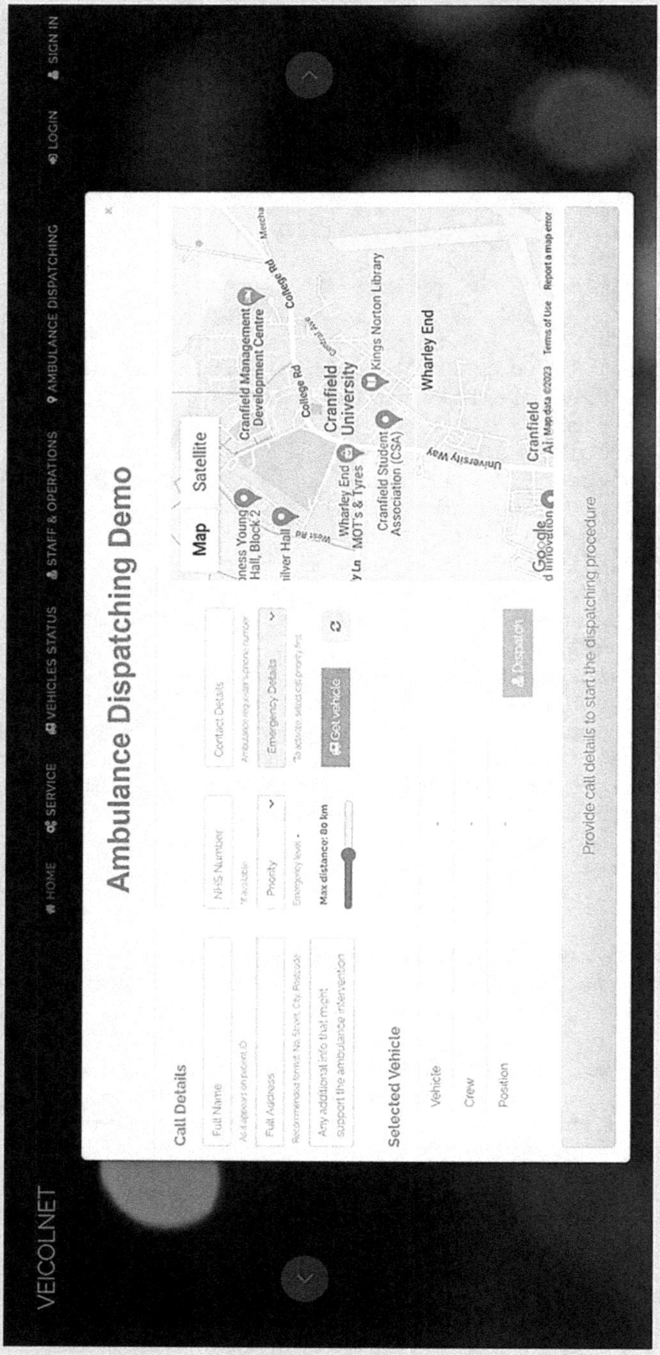

FIGURE D.5 Web app, UI details: Ambulance dispatch, empty form.

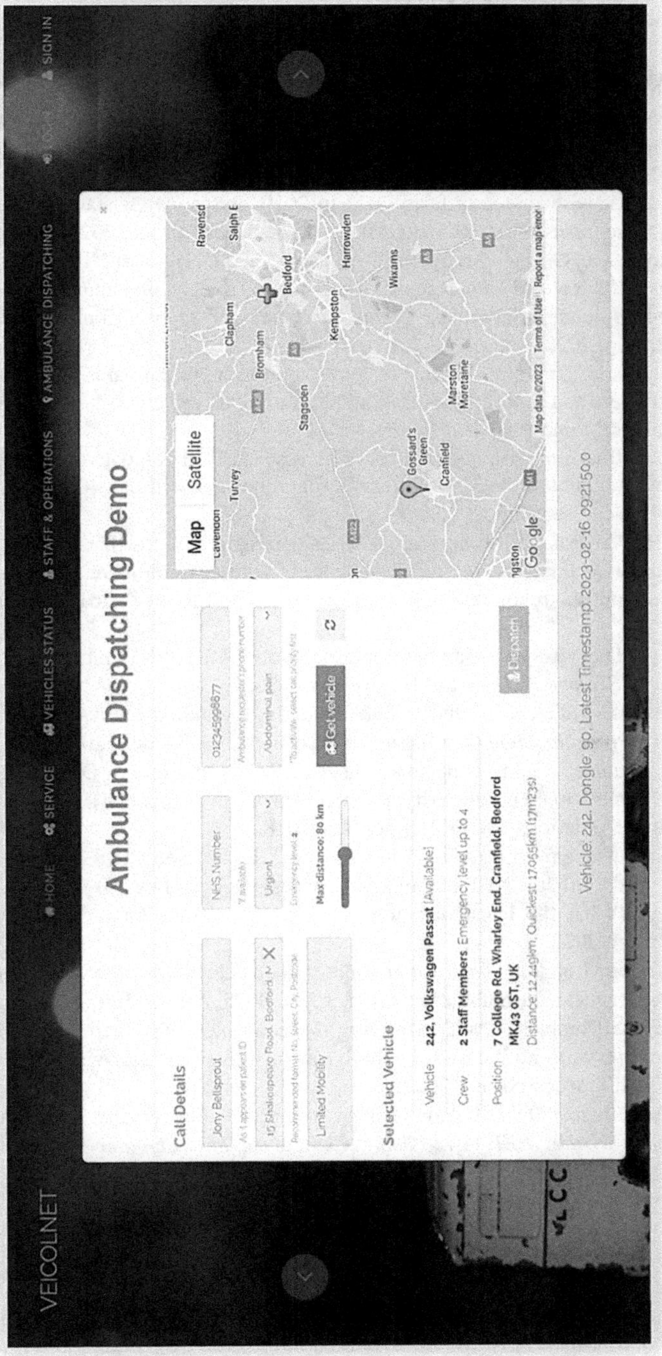

FIGURE D.6 Web app, UI details: Ambulance dispatch, selected vehicle.

Bibliography

1. M. Patel, J. Shangkuan, and C. Thomas, "What's new with the Internet of Things?" https://www.mckinsey.com/industries/semiconductors/our-insights/whats-new-with-the-internet-of-things, 2017, accessed: Jan. 2022.
2. L. S. Vailshery, "Number of Internet of Things (IoT) connected devices worldwide from 2019 to 2021, with forecasts from 2022 to 2030," https://www.statista.com/statistics/1183457/iot-connected-devices-worldwide, 2022, accessed: Jan. 2022.
3. M. Shafiq, Z. Gu, O. Cheikhrouhou, W. Alhakami, and H. Hamam, "The rise of "Internet of Things": Review and open research issues related to detection and prevention of IoT-based security attacks," *Wireless Communications and Mobile Computing*, vol. 2022, pp. 1–12, Aug. 2022.
4. J. M. Anderson, N. Kalra, K. D. Stanley, P. Sorensen, C. Samaras, O. A. Oluwatola, R. Corporation, and S. R. Transportation, "Autonomous vehicle technology: A guide for policymakers," *Rand Corporation*, 2014.
5. P. Nelson, "Just one autonomous car will use 4,000 gb of data/day," https://www.networkworld.com/article/3147892/one-autonomous-car-will-use-4000-gb-of-dataday.html, 2016, accessed: Jan. 2022.
6. P. Gao, H.-W. Kaas, D. Mohr, and D. Wee, "Disruptive trends that will transform the auto industry," https://www.mckinsey.com/industries/automotive-and-assembly/our-insights/disruptive-trends-that-will-transform-the-auto-industry, 2016, accessed: Jan. 2022.
7. J. Rowley, "The wisdom hierarchy: Representations of the DIKW hierarchy," *Journal of Information Science*, vol. 33, pp. 163–180, Apr. 2007.
8. L. F. Rahman, T. Ozcelebi, and J. Lukkien, "Understanding iot systems: A life cycle approach," *Procedia Computer Science*, vol. 130, pp. 1057–1062, 2018.
9. NHS, "Ambulance quality indicators data 2022-23," NHS, Report, Dec. 2022.
10. P. Carter, "Operational productivity and performance in English NHS ambulance trusts: Unwarranted variations," NHS, Report, Sep. 2018.
11. NHS, "Ambulance response programme review," Report, Sep. 2018.
12. National Audit Office, "NHS ambulance services," Report, Jan. 2017.
13. J. Plewes, "What the latest data tell us about ambulance handover delays," 2022, accessed: Apr. 2022.
14. King's College London, "Data for ambulance dispatch," https://www.kcl.ac.uk/policy-institute/assets/data-for-ambulance-dispatch.pdf, 2022, accessed: Apr. 2022.
15. McKinsey & Company, "What's driving the connected car," https://www.mckinsey.com/industries/automotive-and-assembly/our-insights/whats-driving-the-connected-car, Sept. 2014, accessed: Jul. 2022.
16. B. M. Abdelhafiz, "Distributed database using sharding database architecture," in *Proceedings of the IEEE Asia-Pacific Conference on Computer Science and Data Engineering (CSDE)*. Gold Coast, Australia: IEEE, Dec. 2020, pp. 1–17.
17. G. D. Abowd, A. K. Dey, P. J. Brown, N. Davies, M. Smith, and P. Steggles, "Towards a better understanding of context and context-awareness," in *Proceedings of the 1st international symposium on Handheld and Ubiquitous Computing (HUC'99)*. Berlin, Germany: Springer, Sep. 1999, pp. 304–307.
18. D. Ajtay and M. Weilenmann, "Static and dynamic instantaneous emission modelling," *International Journal of Environment and Pollution*, vol. 22, p. 226, 2004.

19. A. Alarilla, M. Stafford, E. Coughlan, J. Keith, and C. Tallack, "The Health Foundation: Why have ambulance waiting times been getting worse?" https://www.health.org.uk/publications/long-reads/why-have-ambulance-waiting-times-been-getting-worse, 2022, accessed: Jun. 2022.
20. H. K. Aliwarga, A. H. Satriatama, and B. F. A. Pratama, "Performance comparison of fleet management system using IoT node device based on MQTT and HTTP protocol," in *AIP Conference Proceedings*, Surakarta, Indonesia, 2020, p. 020009.
21. A. Aljaafreh, M. Khalel, I. Al-Fraheed, K. Almarahleh, R. Al-Shwaabkeh, S. Al-Etawi, and W. Shaqareen, "Vehicular data acquisition system for fleet management automation," in *Proceedings of the IEEE International Conference on Vehicular Electronics and Safety*. Beijing, China: IEEE, Jul. 2011, pp. 130–133.
22. I. M. Almomani, N. Y. Alkhalil, E. M. Ahmad, and R. M. Jodeh, "Ubiquitous GPS vehicle tracking and management system," in *Proceedings of the IEEE Jordan Conference on Applied Electrical Engineering and Computing Technologies (AEECT)*. Amman, Jordan: IEEE, Dec. 2011, pp. 1–6.
23. S. Z. K. Al-Shaqsi, "Response time as a sole performance indicator in EMS: Pitfalls and solutions," *Open Access Emergency Medicine: OAEM*, vol. 2, p. 1, 2010.
24. M. Amorim, S. Ferreira, and A. Couto, "Emergency medical service response: Analyzing vehicle dispatching rules," *Transportation Research Record: Journal of the Transportation Research Board*, vol. 2672, pp. 10–21, Dec. 2018.
25. Android, "Develop android apps with kotlin," https://developer.android.com/kotlin, 2022a, accessed: Sep. 2022.
26. Android, "Guide to app architecture," https://developer.android.com/topic/architecture, 2022b, accessed: Sep. 2022.
27. Android, "Navigation," https://developer.android.com/guide/navigation, 2022c, accessed: Sep. 2022.
28. Apache Kafka, https://kafka.apache.org, 2022, accessed: Jun. 2022.
29. Apache Software Foundation, "Apache tomcat, official documentation," https://tomcat.apache.org/tomcat-8.5-doc, 2022, accessed: Dec. 2022.
30. APAI-CRVS, "Define system requirements," http://www.crvs-dgb.org/en/activities/analysis-and-design/8-define-system-requirements, 2014, accessed: Feb. 2022.
31. Axios, "Getting started, axios docs," https://axios-http.com/docs/intro, 2022, accessed: Jun. 2022.
32. S.-H. Back, J.-J. Kim, M.-J. Kim, H.-S. Kim, Y.-S. Park, and J.-W. Jang, "Implementation of the vehicle black box using external sensor and networks," in Tai-hoon Kim, Alan Chin-Chen Chang, MingChu Li, Chunming Rong, Charalampos Z. Patrikakis, and Dominik Ślęzak (eds.), *Communications in Computer and Information Science*. Springer, 2010, vol. 119, pp. 217–226.
33. J. Backman, J. Vare, K. Framling, M. Madhikermi, and O. Nykanen, "IoT-based interoperability framework for asset and fleet management," in *Proceedings of the IEEE 21st International Conference on Emerging Technologies and Factory Automation (ETFA)*. Berlin, Germany: IEEE, Sep. 2016, pp. 1–4.
34. A. E. Bagaskara, S. Setyorini, and A. A. Wardana, "Performance analysis of message broker for communication in fog computing," in *Proceedings of the 12th International Conference on Information Technology and Electrical Engineering (ICITEE)*, Yogyakarta, Indonesia, 2020, pp. 98–103.
35. T. Berners-Lee, R. Fielding, and H. Frystyk, "RFC1945: Hypertext Transfer Protocol–HTTP/1.0," 1996.

36. G. Bevan and R. Hamblin, "Hitting and missing targets by ambulance services for emergency calls: effects of different systems of performance measurement within the UK," *Journal of the Royal Statistical Society: Series A (Statistics in Society)*, vol. 172, no. 1, pp. 161–190, 2009.
37. B. Bibeault, A. D. Rosa, and Y. Katz, *jQuery in Action*. Simon and Schuster, 2015.
38. H. Billhardt, M. Lujak, V. Sanchez-Brunete, A. Fernandez, and S. Ossowski, "Dynamic coordination of ambulances for emergency medical assistance services," *Knowledge-Based Systems*, vol. 70, pp. 268–280, Nov. 2014.
39. P. S. Bist, "Monitoring your fleets on the go with RTDFMS," Nov. 2020.
40. A. Blandford and B. W. Wong, "Situation awareness in emergency medical dispatch," *International Journal of Human-Computer Studies*, vol. 61, no. 4, pp. 421–452, 2004.
41. BMW Group, "DriveNow becomes wholly-owned subsidiary of BMW Group," https://www.press.bmwgroup.com/global/article/detail/T0278312EN/drivenow-becomes-wholly-owned-subsidiary-of-bmw-group, 2022, accessed: Jul. 2022.
42. A. Bogie, "IAAF and FIGIEFA delighted with provisional type-approval legislation," https://www.tyrepress.com/2018/01/iaaf-and-figiefa-delighted-with-provisional-type-approval-legislation, Jan. 2018, accessed: Jul. 2022.
43. K. Bohm and L. Kurland, "The accuracy of medical dispatch-a systematic review," *Scandinavian Journal of Trauma, Resuscitation and Emergency Medicine*, vol. 26, pp. 1–10, 2018.
44. Bootstrap, "Build fast, responsive sites with Bootstrap," https://getbootstrap.com, 2022, accessed: Sep. 2022.
45. K. Brennan, *A Guide to the Business Analysis Body of Knowledger*. Iiba, 2019.
46. P. J. Brown, "The stick-e document: a framework for creating context-aware applications," in *Proceedings of EP'96*, Palo Alto. Also published in it EP–odd, Jan. 1996, pp. 182–196 [Online]. Available: http://www.cs.kent.ac.uk/pubs/1996/396
47. Bureau of Health Information, "Measuring and reporting performance of NSW ambulance services," Bureau of Health Information, Report, Jun. 2017.
48. R. K. Cheung, D. D. Hang, and N. Shi, "A labeling method for dynamic driver-task assignment with uncertain task durations," *Operations Research Letters*, vol. 33, pp. 411–420, Jul. 2005.
49. R. Chow and K. J. Vicente, "A field study of emergency ambulance dispatching: Implications for decision support," in *Proceedings of the Human Factors and Ergonomics Society Annual Meeting*, vol. 46, no. 3. Baltimore, MD: SAGE Publications, 2002, pp. 313–317.
50. J. Clawson, C. H. Olola, A. Heward, G. Scott, and B. Patterson, "Accuracy of emergency medical dispatchers' subjective ability to identify when higher dispatch levels are warranted over a Medical Priority Dispatch System automated protocol's recommended coding based on paramedic outcome data," *Emergency Medicine Journal*, vol. 24, no. 8, pp. 560–563, 2007.
51. A. Clayton, "Requirements specification - ambulance data set directions 2022," Department of Health and Social Care, Report, Jun. 2022.
52. E. F. Codd, "A relational model of data for large shared data banks," *Communications of the ACM*, vol. 13, pp. 377–387, Jun. 1970.
53. Confluent, "Never at rest - IoT and data streaming at British gas connected homes," https://www.confluent.io/resources/kafka-summit-2016/never-rest-iot-data-streaming-british-gas-connected-homes, 2022, accessed: Jun. 2022.
54. L. Coslovich, R. Pesenti, and W. Ukovich, "Minimizing fleet operating costs for a container transportation company," *European Journal of Operational Research*, vol. 171, pp. 776–786, Jun. 2006.

55. DataSF, "Law enforcement dispatched calls for service – Dataset Explainers," https://sfdigitalservices.gitbook.io/dataset-explainers/law-enforcement-dispatched-calls-for-service, 2023, accessed: Sep. 2023.
56. DataSF, "San Francisco police districts," https://data.sfgov.org/Public-Safety/Current-Police-Districts/wkhw-cjsf, 2023, accessed: Sep. 2023.
57. E. Demir, T. Bektas, and G. Laporte, "A comparative analysis of several vehicle emission models for road freight transportation," *Transportation Research Part D: Transport and Environment*, vol. 16, pp. 347–357, Jul. 2011.
58. M. J. F. de Oliveira and L. N. P. Toscano, "An integrated emergency care delivery system for major events," *Operations Research for Health Care*, vol. 17, pp. 16–27, Jun. 2018.
59. J. Desjardins, "How much data is generated each day?" https://www.visualcapitalist.com/how-much-data-is-generated-each-day, 2019, accessed: Apr. 2022.
60. A. K. Dey, "Understanding and using context," *Personal and Ubiquitous Computing*, vol. 5, pp. 4–7, Feb. 2001.
61. H. Dhaduk, "6 emerging fleet management technologies you need to know," https://www.simform.com/blog/emerging-technologies-fleet-management, 2021, accessed: Feb. 2022.
62. A. Drake, A. Pollitt, L. Smith, and E. Sklar, "Data awareness for sending help (DASH): policy opportunities & challenges," *Emergency Medicine Journal*, vol. 36, pp. e12.2–e12.1, 2019.
63. S. Dunnett, J. Leigh, and L. Jackson, "Optimising police dispatch for incident response in real time," *Journal of the Operational Research Society*, vol. 70, pp. 269–279, Feb. 2019.
64. M. H. Eiza and Q. Ni, "Driving with sharks: Rethinking connected vehicles with vehicle cybersecurity," *IEEE Vehicular Technology Magazine*, vol. 12, pp. 45–51, Jun. 2017.
65. S. Enayati, M. E. Mayorga, H. K. Rajagopalan, and C. Saydam, "Real-time ambulance redeployment approach to improve service coverage with fair and restricted workload for EMS providers," *Omega*, vol. 79, pp. 67–80, Sep. 2018.
66. Express.js, "Express — Fast, unopinionated, minimalist web framework for Node.js," https://expressjs.com, 2022, accessed: Jun. 2022.
67. M. Falch, "OBD2 explained – A simple intro," https://www.csselectronics.com/pages/obd2-explained-simple-intro, Apr. 2022, accessed: Jul. 2022.
68. E. Fernandes, "Fleet management system market size worth $42.75 billion, globally, by 2028 at 10.82% CAGR: Verified market research," https://www.globenewswire.com/news-release/2021/11/24/2340703/0/en/Fleet-Management-System-Market-size-worth-42-75-Billion-Globally-by-2028-at-10-82-CAGR-Verified-Market-Research.html, 2021, accessed: Feb. 2022.
69. A. Fink and T. Reiners, "Modeling and solving the short-term car rental logistics problem," *Transportation Research Part E: Logistics and Transportation Review*, vol. 42, pp. 272–292, Jul. 2006.
70. M. Freedman and A. Sewrathan, "TimescaleDB vs. InfluxDB: Purpose built differently for time-series data," https://www.timescale.com/blog/timescaledb-vs-influxdb-for-time-series-data-timescale-influx-sql-nosql-36489299877, 2020, accessed: May 2022.
71. R. Fulton and V. Roy, "Requirements - Writing requirements," in *Airborne Electronic Hardware Design Assurance: A Practitioner's Guide to RTCA/DO-254*. CRC Press, 2017, pp. 89–93.
72. FVT, "PHEM, passenger car and heavy duty emission model," https://www.fvt.at/em/en/simulation/phem.html, 2022, accessed: May 2022.

73. S. Gadam, "What is the iterative and incremental development process?" https://blog.logrocket.com/product-management/what-is-iterative-incremental-development-process-examples, 2023, accessed: Feb. 2023.
74. V. J. Gawron, C. G. Drury, R. J. Fairbanks, and R. C. Berger, "Medical error and human factors engineering: where are we now?" *American Journal of Medical Quality*, vol. 21, no. 1, pp. 57–67, 2006.
75. Geotab, "Geotab: Fleet management solutions," https://www.geotab.com/uk, 2022, accessed: Jun. 2022.
76. J. Gerhardt-Powals, "Cognitive engineering principles for enhancing human-computer performance," *International Journal of Human-Computer Interaction*, vol. 8, no. 2, pp. 189–211, 1996.
77. Github - Google/Volley, https://github.com/google/volley, 2022, accessed: Jun. 2022.
78. Google, "Google maps platform APIs by platform," https://developers.google.com/maps/apis-by-platform, 2022, accessed: Jun. 2022.
79. Y. Guo, "On the expansion of on-board diagnostics (OBD) to electric propulsion systems in battery electric vehicles," *SAE International Journal of Advances and Current Practices in Mobility*, vol. 4, pp. 157–166, Apr. 2021.
80. T. Haerder and A. Reuter, "Principles of transaction-oriented database recovery," *ACM Computing Surveys*, vol. 15, pp. 287–317, Dec. 1983.
81. A. Haghani and S. Yang, "Dynamic fleet management," in Vasileios Zeimpekis, Christos D. Tarantilis, George M. Giaglis, and Ioannis Minis (eds.), *Real-Time Emergency Response Fleet Deployment: Concepts, Systems, Simulation & Case Studies*, Springer, 2007, pp. 133–162.
82. N. Haines, "Beginning Ubuntu for Windows and Mac users," in *The Future of Ubuntu.* Apress, 2015, pp. 205–209.
83. M. M. Hassan, M. R. Hassan, V. H. C. de Albuquerque, and W. Pedrycz, "Soft computing for intelligent edge computing," *Applied Soft Computing*, vol. 128, p. 109628, Oct. 2022.
84. J. Hayes, A. Moore, G. Benwell, and B. Wong, "Identifying decision complexity factors in ambulance command and control," in *Proceedings of the Human Factors and Ergonomics Society Annual Meeting*, vol. 49, no. 3. Orlando, FL: SAGE Publications, 2005, pp. 234–237.
85. Hibernate, "Hibernate—Everything data," https://hibernate.org, 2022, accessed: Sep. 2022.
86. HiveMQ, "HiveMQ clusters," www.hivemq.com/docs/hivemq/4.13/user-guide/cluster.html, 2022b, accessed: Jun 2022.
87. HiveMQ, "Publish subscribe - MQTT essentials: Part 2," https://www.hivemq.com/blog/mqtt-essentials-part2-publish-subscribe, 2022a, accessed: Jun. 2022.
88. B. Hjørland, "Data (with big data and database semantics)," *KO Knowledge Organization*, vol. 45, no. 8, pp. 685–708, 2019.
89. A. Hoff, H. Andersson, M. Christiansen, G. Hasle, and A. Lokketangen, "Industrial aspects and literature survey: Fleet composition and routing," *Computers Operations Research*, vol. 37, pp. 2041–2061, Dec. 2010.
90. E. Hollnagel, "Cognitive ergonomics: it's all in the mind," *Ergonomics*, vol. 40, no. 10, pp. 1170–1182, 1997.
91. L. Hoory and C. Bottorff, "Agile vs. waterfall: Which project management methodology is best for you?" https://www.forbes.com/advisor/business/agile-vs-waterfall-methodology, 2022, accessed: Apr. 2022.
92. Human Service Solutions, "Mobile website vs. mobile app: Which is best for your organization?" https://www.hswsolutions.com/services/mobile-web-development/mobile-website-vs-apps, 2022, accessed: Apr. 2022.

93. E. T. Hvannberg, E. L.-C. Law, and M. K. Lérusdóttir, "Heuristic evaluation: Comparing ways of finding and reporting usability problems," *Interacting with Computers*, vol. 19, no. 2, pp. 225–240, 2007.
94. IBM, "Building native add-ons," https://www.ibm.com/docs/en/sdk-nodejs-zos/16.0?topic=started-building-native-add-ons, Apr. 2022, accessed: Jul. 2022.
95. IBM, "What is software testing?" https://www.ibm.com/topics/software-testing, 2022, accessed: Dec. 2022.
96. ICURe, "ICURe programme," https://www.icureprogramme.com, n.d., accessed: Jul. 2022.
97. influxdata, "InfluxDB times series data platform," https://www.influxdata.com, 2022, accessed: May 2022.
98. Institute of Asset Management, "Asset management - An anatomy, version 3," Tech. Rep., Dec. 2015.
99. Institute of Electrical and Electronics Engineers, "Asset management - overview, principles and terminology," Standard, Jan. 2014.
100. Institute of Electrical and Electronics Engineers, "IEEE standard glossary of software engineering terminology," Standard, Sep. 1990.
101. Intellias, "Big data for connected car platforms: What's under the hood?" https://intellias.com/big-data-for-connected-car-platforms, 2020, accessed: Feb. 2022.
102. ISO/IEC/IEEE 24765:2010(E), "ISO/IEC/IEEE international standard - systems and software engineering – vocabulary," pp. 1–418, Dec. 2010.
103. V. Jaiprakash, "Survey on MongoDB: An open-source document database," *International Journal of Advanced Research in Engineering and Technology*, vol. 6, pp. 1–11, 2015.
104. N. Jatana, S. Puri, M. Ahuja, I. Kathuria, and D. Gosain, "A survey and comparison of relational and non-relational database," *International Journal of Engineering Research Technology*, vol. 1, pp. 1–5, 2012.
105. JavaScript Obfuscator, "javascript-obfuscator," *GitHub Repository*, https://github.com/javascript-obfuscator, 2022, accessed: Jul. 2022.
106. C. Kaner, "The power of what if… and nine ways to fuel your imagination: Cem kaner on scenario testing," *Software Testing and Quality Engineering*, vol. 5, pp. 16–22, 2003.
107. Kernel Organization, "Socketcan - controller area network (official documentation)," https://www.kernel.org/doc/html/latest/networking/can.html, 2022, accessed: Jul. 2022.
108. R. Kiefer, "TimescaleDB vs. PostgreSQL for time-series: 20x higher inserts, 2000x faster deletes, 1.2x-14,000x faster queries," https://www.timescale.com/blog/timescaledb-vs-6a696248104e, 2017, accessed: May 2022.
109. D. Klinedinst and C. King, "On board diagnostics: Risks and vulnerabilities of the connected vehicle," CERT Coordination Center, Tech. Rep, 2016.
110. V. Knight, P. Harper, and L. Smith, "Ambulance allocation for maximal survival with heterogeneous outcome measures," *Omega*, vol. 40, pp. 918–926, Dec. 2012.
111. E. Knowles, L. Bishop-Edwards, and A. O'Cathain, "Exploring variation in how ambulance services address non-conveyance: A qualitative interview study," *BMJ Open*, vol. 8, p. e024228, Nov. 2018.
112. M. Lachgar, H. Benouda, and S. Elfirdoussi, "Android REST APIs: Volley vs retrofit," in *International Symposium on Advanced Electrical and Communication Technologies (ISAECT)*, Rabat, Morocco, 2018, pp. 1–6.
113. F. Lardinois, "Google makes kotlin a first-class language for writing android apps," https://techcrunch.com/2017/05/17/google-makes-kotlin-afirst-class-language-for-writing-android-apps, 2017, accessed: Sep. 2022.
114. N. R. Levick and J. S. Metroems, "An optimal solution for enhancing ambulance safety: Implementing a driver performance feedback and monitoring device in ground emergency medical service vehicles," in *Annual proceedings. Association for the Advancement of Automotive Medicine*, vol. 49. Boston, MA: Elsevier, 2005, pp. 35–50.

115. L. Y. O. Li and Z. Fu, "The school bus routing problem: a case study," *Journal of the Operational Research Society*, vol. 53, pp. 552–558, May 2002.
116. K. L. Lim, J. Whitehead, D. Jia, and Z. Zheng, "State of data platforms for connected vehicles and infrastructures," *Communications in Transportation Research*, vol. 1, p. 100013, Dec. 2021.
117. F. M. S. Lima, D. S. Pereira, S. V. Conceicao, and N. T. R. Nunes, "A mixed load capacitated rural school bus routing problem with heterogeneous fleet: Algorithms for the Brazilian context," *Expert Systems with Applications*, vol. 56, pp. 320–334, Sep. 2016.
118. C. Lin, C.-C. Li, S.-H. Yang, S.-H. Lin, and C.-Y. Lin, "Development of on-line diagnostics and real time early warning system for vehicles," in *Sensors for Industry Conference*. Houston, TX: IEEE, Feb. 2005, pp. 45–51.
119. M. L. Manna, L. Treccozzi, P. Perazzo, S. Saponara, and G. Dini, "Performance evaluation of attribute-based encryption in automotive embedded platform for secure software over-the-air update," *Sensors*, vol. 21, p. 515, Jan. 2021.
120. I. Markov, S. Varone, and M. Bierlaire, "Integrating a heterogeneous fixed fleet and a flexible assignment of destination depots in the waste collection VRP with intermediate facilities," *Transportation Research Part B: Methodological*, vol. 84, pp. 256–273, Feb. 2016.
121. J. L. Martin, B. J. Norris, E. Murphy, and J. A. Crowe, "Medical device development: The challenge for ergonomics," *Applied Ergonomics*, vol. 39, no. 3, pp. 271–283, 2008.
122. J. E. Meseguer, C. K. Toh, C. T. Calafate, J. C. Cano, and P. Manzoni, "Driving styles: a mobile platform for driving styles and fuel consumption characterization," *Journal of Communications and Networks*, vol. 19, pp. 162–168, Apr. 2017.
123. MICRA, "MICRA Gateway," https://micragateway.org, n.d., accessed: Jul. 2022.
124. G. Milette, "Reducing code using kotlin," https://www.tripadvisor.com/engineering/reducing-code-using-kotlin, 2018, accessed: Sep. 2022.
125. A. M. Mirza, M. N. A. Khan, R. A. Wagan, M. B. Laghari, M. Ashraf, M. Akram, and M. Bilal, "Contextdrive: Towards a functional scenario-based testing framework for context-aware applications," *IEEE Access*, vol. 9, pp. 80478–80490, 2021.
126. B. Mishra and A. Kertesz, "The use of MQTT in M2M and IoT systems: A survey," *IEEE Access*, vol. 8, pp. 201071–201086, 2020.
127. A. Mortaro, D. Pascu, T. Zerman, E. Vallaperta, A. Schönsberg, S. Tardivo, S. Pancheri, G. Romano, and F. Moretti, "The role of the emergency medical dispatch centre (EMDC) and prehospital emergency care safety: results from an incident report (IR) system," *Canadian Journal of Emergency Medicine*, vol. 17, no. 4, pp. 411–419, 2015.
128. Mozilla, "Connection management in HTTP/1.x," https://developer.mozilla.org/en-US/docs/Web/HTTP/Connection_management_in_HTTP_1.x, 2022, accessed: Jun. 2022.
129. J. B. Myers, C. M. Slovis, M. Eckstein, J. M. Goodloe, S. M. Isaacs, J. R. Loflin, C. C. Mechem, N. J. Richmond, and P. E. Pepe, "Evidence-based performance measures for emergency medical services systems: a model for expanded EMS benchmarking: a statement developed by the 2007 consortium us metropolitan municipalities' EMS medical directors," *Prehospital Emergency Care*, vol. 12, no. 2, pp. 141–151, 2008.
130. NAEI, "NAEI, UK national atmospheric emissions inventory," https://naei.beis.gov.uk/data, 2022, accessed: May 2022.
131. N. Naik, "Choice of effective messaging protocols for IoT systems: MQTT, CoAP, AMQP and HTTP," in *Proceedings of the IEEE International Systems Engineering Symposium (ISSE)*. Vienna, Austria: IEEE, Oct. 2017, pp. 1–7.
132. ngrok, "ngrok — All your traffic. One gateway," https://ngrok.com, 2022, accessed: Dec. 2022.

133. NHS, "A guide for healthcare professionals: Requesting ambulance transport," https://www.nwas.nhs.uk/news/healthcare-professional-requests-for-ambulance-transport, 2019, accessed: Apr. 2022.
134. NHS, "Publication of 2016 to 2017 NHS trust accounting returns data," Department of Health and Social Care, Report, Aug. 2017.
135. NHS, "Record number of NHS ambulance call outs for life-threatening conditions in December, despite jump in omicron absences," https://www.england.nhs.uk/2022/01/record-number-of-nhs-ambulance-call-outs-for-life-threatening-conditions-in-december, 2022, accessed: Apr. 2022.
136. S. Nidhra and J. Dondeti, "Black box and white box testing techniques-a literature review," *International Journal of Embedded Systems and Applications (IJESA)*, vol. 2, no. 2, pp. 29–50, 2012.
137. Node.js, "Don't block the event loop (or the worker pool)," https://nodejs.org/en/docs/guides/dont-block-the-event-loop, 2022, accessed: Jun. 2022.
138. M. Noel, "How car makers are trying to close the OBD port and keep control over the car data?," https://www.linkedin.com/pulse/how-car-makers-trying-close-obd-port-keep-control-over-matthieu-noel/, Oct. 2017, accessed: Jul. 2022.
139. E. Nordstrom, "Achieving optimal query performance with a distributed time-series database on PostgreSQL," https://www.timescale.com/blog/achieving-optimal-query-performance-with-a-distributed-time-series-database-on-postgresql, 2019, accessed: May 2022.
140. R. O'Hara, M. Johnson, E. Hirst, A. Weyman, D. Shaw, P. Mortimer, C. Newman, M. Storey, J. Turner, S. Mason et al., "Decision making and safety in ambulance service transitions," *Emergency Medicine Journal*, vol. 32, no. 5, p. e2, 2015.
141. B. B. Oliveira, M. A. Carravilla, J. F. Oliveira, and F. M. Toledo, "A relax-and-fix-based algorithm for the vehicle-reservation assignment problem in a car rental company," *European Journal of Operational Research*, vol. 237, pp. 729–737, Sep. 2014.
142. Openbase, "Streaming-worker and streaming-worker-SDK, official documentation," https://openbase.com/js/streaming-worker-sdk/documentation, 2022, accessed: Jul. 2022.
143. J. Orasanu and U. Fischer, "Improving healthcare communication: Lessons from the flightdeck," in Christopher P. Nemeth (ed.), *Improving Healthcare Team Communication*. CRC Press, 2017, pp. 23–45.
144. Otonomo Platform, "Otonomo connected-car data platform," https://otonomo.io/platform, 2022, accessed: Jul. 2022.
145. Outils OBD Facile, "OBD diagnostics and automotive tools," https://www.outilsobdfacile.com, 2022, accessed: Jul. 2022.
146. J. Park, H. Tae, and B.-I. Kim, "A post-improvement procedure for the mixed load school bus routing problem," *European Journal of Operational Research*, vol. 217, pp. 204–213, Feb. 2012.
147. E. Peters, R. de Matta, and W. Boe, "Short-term work scheduling with job assignment flexibility for a multi-fleet transport system," *European Journal of Operational Research*, vol. 180, pp. 82–98, Jul. 2007.
148. PostGIS, "About PostGIS," https://postgis.net, 2022, accessed: May 2022.
149. PostgreSQL, "PostgreSQL documentation: pgcrypto," https://www.postgresql.org/docs/current/pgcrypto.html, 2022, accessed: May 2022.
150. K. C. Preusse and C. Gipson, "Dispatching information in 911 teams: A case study," in *Proceedings of the Human Factors and Ergonomics Society Annual Meeting*, vol. 60, no. 1. Washington, DC: SAGE Publications, 2016, pp. 117–121.

151. PRLOG, "Project alloyed consortium to advance development and commercialisation of UK connected autonomous vehicle technologies," https://www.prlog.org/12645255-project-alloyed-consortium-to-advance-development-and-commercialisation-of-uk-connected-autonomous-vehicle-technologies.html, 2022, accessed: Jan. 2022.
152. A. Quaile, "What's next for England's ambulance services?" *Journal of Paramedic Practice*, vol. 9, pp. 443–444, Oct. 2017.
153. J. Radcliffe and G. Heath, "Ambulance calls and cancellations: policy and implementation issues," *International Journal of Public Sector Management*, vol. 22, no. 5, pp. 410–422, 2009.
154. M. Rahman, "Emergency medical responders and physicians: Diagnostics, decision making and medical management in high stakes situations," in *Proceedings of the Symposium on Human Factors and Ergonomics in Health Care*, Baltimore, MD, 2012, pp. 14–19.
155. R. Rak, "Vehicle identification number – anatomy of error occurrence," *Journal of Physics: Conference Series*, vol. 1303, p. 012146, Aug. 2019.
156. Raspberry pi, "Raspberry pi 3 model b+," https://www.raspberrypi.com/products/raspberry-pi-3-model-b-plus, 2022, accessed: Jul. 2022.
157. M. J. Reilly, "Accuracy of a priority medical dispatch system in dispatching cardiac emergencies in a suburban community," *Prehospital and Disaster Medicine*, vol. 21, no. 2, pp. 77–81, 2006.
158. ReportsandData, "Global vehicle telematics market size to reach USD 779.98 billion in 2030," https://www.reportsanddata.com/press-release/global-vehicle-telematics-market, Jun. 2022, accessed: Jul. 2022.
159. E. Russell, "9 things to know about Google's maps data," https://cloud.google.com/blog/products/maps-platform/9-things-know-about-googles-maps-data-beyond-map, 2023, accessed: Mar. 2023.
160. R. Salazar-Cabrera, A. P. D. L. Cruz, and J. M. M. Molina, "Fleet management and control system from intelligent transportation systems perspective," in *Proceedings of the 2nd Latin American Conference on Intelligent Transportation Systems (ITS LATAM)*. Bogota, Colombia: IEEE, Mar. 2019, pp. 1–7.
161. Sam, "How to use a LCD screen/display with raspberry pi," *Core Electronics*, https://core-electronics.com.au/guides/how-to-use-a-screen-display-with-raspberry-pi, 2022, accessed: Jul. 2022.
162. B. Schilit and M. Theimer, "Disseminating active map information to mobile hosts," *IEEE Network*, vol. 8, pp. 22–32, Sep. 1994.
163. B. Shaik, "User management and securing databases," in *PostgreSQL Configuration*. Apress, 2020, pp. 61–92.
164. K. Shin and P. Ramanathan, "Real-time computing: A new discipline of computer science and engineering," *Proceedings of the IEEE*, vol. 82, pp. 6–24, 1994.
165. A. Singh, "Android dev summit '18 - everything you need to know," https://medium.com/androidiots/android-dev-summit-18-everything-youneed-to-know-fc5583cd7f90, 2022, accessed: Sep. 2022.
166. SK Pang Electronics, "Pican2 - can interface for raspberry pi with SMPS," https://www.skpang.co.uk/products/pican2-can-bus-board-for-raspberry-pi-2-3-with-smps, 2022, accessed: Jul. 2022.
167. P. Smith, "On the unintended consequences of publishing performance data in the public sector," *International journal of public administration*, vol. 18, no. 2-3, pp. 277–310, 1995.
168. M. Sojka, P. Pisa, M. Petera, O. Spinka, and Z. Hanzalek, *A Comparison of Linux CAN Drivers and Their Applications*. IEEE, Jul. 2010, pp. 18–27.

169. Stack Overflow, "2022 developer survey," https://survey.stackoverflow.co/2022, 2022, accessed: May 2022.
170. Statista, "Largest vehicle hire and leasing companies based on the number of vehicles in their fleet in the united kingdom (UK) as of November 2022," https://www.statista.com/statistics/321436/vehicle-hiring-company-fleet-sizes-in-the-united-kingdom, 2023, accessed: Feb. 2022.
171. D. Stojanovic, B. Predic, I. Antolovic, and S. Dordevic-Kajan, "Web information system for transport telematics and fleet management," in *9th International Conference on Telecommunication in Modern Satellite, Cable, and Broadcasting Services*. Nis, Serbia: IEEE, Oct. 2009, pp. 314–317.
172. Z. Szalay, Z. Kanya, L. Lengyel, P. Ekler, T. Ujj, T. Balogh, and H. Charaf, "ICT in road vehicles – reliable vehicle sensor information from OBD versus CAN," in *Proceedings of the International Conference on Models and Technologies for Intelligent Transportation Systems (PMT-ITS)*. Budapest, Hungary: IEEE, Jun. 2015, pp. 469–476.
173. P. Taylor, "Mobile operating systems' market share worldwide from 1st quarter 2009 to 4th quarter 2022," https://www.statista.com/statistics/272698/global-market-share-held-by-mobile-operating-systems-since-2009, 2023, accessed: Sep. 2022.
174. TimescaleDB, "Hypertable and chunk architecture," https://docs.timescale.com/timescaledb/latest/overview/core-concepts/hypertables-and-chunks/hypertable-architecture, 2022, accessed: May 2022.
175. Time Series Benchmark Suite (TSBS), https://github.com/timescale/tsbs, 2022, accessed: May 2022.
176. R. Tönjes, P. Barnaghi, M. Ali, A. Mileo, M. Hauswirth, F. Ganz, S. Ganea, B. Kjærgaard, D. Kuemper, S. Nechifor et al., "Real time iot stream processing and large-scale data analytics for smart city applications," in *Poster Session, European Conference on Networks and Communications*, Bologna, Italy, 2014, p. 10.
177. Totalcardiagnostics, "OBD2 communication protocols by manufacturer (PDF)," https://www.totalcardiagnostics.com/files/OBD2_protocols.pdf, 2022, accessed: Jul. 2022.
178. K. Vassakis, E. Petrakis, and I. Kopanakis, "Big data analytics: applications, prospects and challenges," in Georgios Skourletopoulos, George Mastorakis, Constandinos X. Mavromoustakis, Ciprian Dobre, and Evangelos Pallis (eds.), *Mobile Big Data. Lecture Notes on Data Engineering and Communications Technologies*. Springer, 2007, vol. 10, pp. 3–20.
179. Vincario, https://vindecoder.eu, 2022, accessed: Jun. 2022.
180. M. Vivaldini, S. R. I. Pires, and F. B. de Souza, "Improving logistics services through the technology used in fleet management," *Journal of Information Systems and Technology Management*, vol. 9, pp. 541–562, Dec. 2012.
181. A. Volle, "Web application," https://www.britannica.com/topic/Web-application, 2022, accessed: Apr. 2022.
182. W3Techs, "Usage statistics of site elements for websites," https://w3techs.com/technologies/overview/site_element, 2022, accessed: Jun. 2022.
183. C. Wang, "HTTP vs. MQTT: A tale of two IoT protocols," https://www.hivemq.com/blog/mqtt-essentials-part2-publish-subscribe, Nov. 2018, accessed: Jun. 2022.
184. P. Wankhade, "Performance measurement and the UK emergency ambulance service: Unintended consequences of the ambulance response time targets," *International Journal of Public Sector Management*, vol. 24, no. 5, pp. 384–402, 2011.
185. A. Ward, A. Jones, and A. Hopper, "A new location technique for the active office," *IEEE Personal Communications*, vol. 4, pp. 42–47, 1997.
186. Waveshare Electronics, "Sim7600e-h 4g dongle – 4G/LTE HAT for Raspberry Pi," https://www.waveshare.com/sim7600e-h-4g-dongle.htm, 2022, accessed: Jul. 2022.

187. J. Whalen, "Expert systems versus systems for experts: Computer-aided dispatch as a support system in real-world environments," *Cambridge Series on Human Computer Interaction*, pp. 161–183, 1995.
188. D. A. Wiegmann and S. A. Shappell, "Human error perspectives in aviation," *The International Journal of Aviation Psychology*, vol. 11, no. 4, pp. 341–357, 2001.
189. E. T. Wilde, "Do emergency medical system response times matter for health outcomes?" *Health Economics*, vol. 22, pp. 790–806, Jul. 2013.
190. E. Wilde and C. Pautasso, *REST: From Research to Practice*. Springer Science & Business Media, 2011.
191. B. W. Wong, J. Hayes, and T. Moore, "What makes emergency ambulance command and control complex?" in C. Johnson (ed.), *2nd Workshop on Complexity in Design and Engineering, GIST Technical Report G2005-1*, Glasgow, Scotland, 2005.
192. B. W. Wong and A. Blandford, "Analysing ambulance dispatcher decision making: Trialing emergent themes analysis," in *Proceedings of the HF2002 Human Factors Conference: Design for the Whole Person: Integrating Physical, Cognitive and Social Aspects*. Melbourne, Australia: Ergonomics Society of Australia, 2002.
193. Y. Yang, B. Chen, L. Su, and D. Qin, "Research and development of hybrid electric vehicles CAN-bus data monitor and diagnostic system through OBD-II and android-based smartphones," *Advances in Mechanical Engineering*, vol. 5, p. 741240, Jan. 2013.
194. Q. C. H. Z. Yang, Zhifeng Chen and G. Teo, "Enabling seamless kafka async queuing with consumer proxy," https://www.uber.com/en-GB/blog/kafka-async-queuing-with-consumer-proxy, accessed: Jun. 2022.

Index

Pages in *italics* refer to figures, pages in ***bold-italics*** refer to boxes and pages in **bold** refer to tables.

D

E

F

G

H

Y

For Product Safety Concerns and Information please contact our EU
representative GPSR@taylorandfrancis.com
Taylor & Francis Verlag GmbH, Kaufingerstraße 24, 80331 München, Germany

www.ingramcontent.com/pod-product-compliance
Lightning Source LLC
LaVergne TN
LVHW011154030926
847199LV00028B/289

* 9 7 8 1 0 4 1 1 7 2 1 4 7 *